T0290873

Applied Systems Analysis

Advanced Research in Reliability and System Assurance Engineering

Series Editor:
Mangey Ram
Professor, Graphic Era University, Uttarakhand, India

Modeling and Simulation Based Analysis in Reliability Engineering
Edited by Mangey Ram

Reliability Engineering
Theory and Applications
Edited by Ilia Vonta and Mangey Ram

System Reliability Management
Solutions and Technologies
Edited by Adarsh Anand and Mangey Ram

Reliability Engineering
Methods and Applications
Edited by Mangey Ram

Reliability Management and Engineering
Challenges and Future Trends
Edited by Harish Garg and Mangey Ram

Applied Systems Analysis
Science and Art of Solving Real-Life Problems
F. P. Tarasenko

Stochastic Models in Reliability Engineering
Lirong Cui, Ilia Frenkel, and Anatoly Lisnianski

For more information about this series, please visit: https://www.crcpress.com/Advanced-Research-in-Reliability-and-System-Assurance-Engineering/book-series/CRCARRSAE

Applied Systems Analysis

Science and Art of Solving Real-Life Problems

F. P. Tarasenko

CRC Press is an imprint of the
Taylor & Francis Group, an **informa** business

First edition published 2021
by CRC Press
6000 Broken Sound Parkway NW, Suite 300, Boca Raton, FL 33487-2742

and by CRC Press
2 Park Square, Milton Park, Abingdon, Oxon, OX14 4RN

© 2021 Taylor & Francis Group, LLC

Previously published in Russian entitled, "Applied Systems Analysis", by Knorus Publisher, Russia.

CRC Press is an imprint of Taylor & Francis Group, LLC

Reasonable efforts have been made to publish reliable data and information, but the author and publisher cannot assume responsibility for the validity of all materials or the consequences of their use. The authors and publishers have attempted to trace the copyright holders of all material reproduced in this publication and apologize to copyright holders if permission to publish in this form has not been obtained. If any copyright material has not been acknowledged please write and let us know so we may rectify in any future reprint.

Except as permitted under U.S. Copyright Law, no part of this book may be reprinted, reproduced, transmitted, or utilized in any form by any electronic, mechanical, or other means, now known or hereafter invented, including photocopying, microfilming, and recording, or in any information storage or retrieval system, without written permission from the publishers.

For permission to photocopy or use material electronically from this work, access www.copyright.com or contact the Copyright Clearance Center, Inc. (CCC), 222 Rosewood Drive, Danvers, MA 01923, 978-750-8400. For works that are not available on CCC please contact mpkbookspermissions@tandf.co.uk

Trademark notice: Product or corporate names may be trademarks or registered trademarks, and are used only for identification and explanation without intent to infringe.

Library of Congress Cataloging-in-Publication Data
Names: Tarasenko, F. P. (Feliks Petrovich), author.
Title: Applied systems analysis : science and art of solving real-life
problems / F.P. Tarasenko.
Description: Boca Raton : CRC Press, 2020. | Series: Advanced research in
reliability and system assurance engineering | Includes bibliographical
references and index.
Identifiers: LCCN 2020009183 (print) | LCCN 2020009184 (ebook) |
ISBN 9780367472399 (hbk) | ISBN 9781003054290 (ebk)
Subjects: LCSH: System analysis.
Classification: LCC T57.6 .T38 2020 (print) | LCC T57.6 (ebook) |
DDC 620.001/1—dc23
LC record available at https://lccn.loc.gov/2020009183
LC ebook record available at https://lccn.loc.gov/2020009184

ISBN: 978-0-367-47239-9 (hbk)
ISBN: 978-1-003-05429-0 (ebk)

Typeset in Times
by codeMantra

Contents

PART II Systems Practice: Technology of Applied Systems Analysis

PART III Brief Review of Results of Systemology in the 20th Century

Preface

The purpose of education is to help a person realize their potential—not only to become a useful member of the society in which they live, but also for the satisfaction of realizing their individual abilities and talents.

The diversity of the specific needs of society and the variety of individual abilities of people are conformed as a result of specialized higher (and partly secondary) education in relation to professions, concentrating the attention and time of students on developing the special knowledge necessary for implementing particular professional activities.

However, the division of human activities into professions, with their specializations, does not repeal the fact that we all live in a single world in which there are common laws and we interact with nature, of which we are a part, from different sides and with different intentions. This fact leads to the conclusion that to achieve success in whatever we do, we must act in accordance with natural laws. The universal unity of nature (and us in it) is reflected in the concept of the systemic nature of all things.

Professionals became aware of the systematic nature of their activities through the experience of success and failure in solving problems. Within the training framework of even the narrowest specialty, there is always a section outlining specific precautions and rules of work that are conducive to success. Therefore, it is not surprising (although the same can be said differently) that the technology of successful problem-solving, that is, the set of recommended and forbidden operations from posing a problem to solving it, is universal. The professional specificity is in which areas knowledge will be needed to solve problems; however, what and how to do with this knowledge is common to all.

Awareness of this in education has been expressed in the introduction to the curricula of many specialties and courses, such as "systems theory", "systems analysis", and "systems engineering". Alas, this has not yet happened in all specialties, and sometimes remains strongly tied to the specifics of the profession (e.g., "methodology of historiography" in history, "methods of engineering creativity" in technology, "general pathology of human body" in medicine). A large number of textbooks, manuals, and teaching materials for individual courses have been created.

It seems appropriate to supplement this arsenal of systemic knowledge by presenting it in an interdisciplinary and supradisciplinary form applicable to solving *any* real-life problem, regardless of the professional specifics of the problem. This section of modern systemology is called "applied systems analysis".

Such a course is conducted at Tomsk State University in Russia for more than 30 years for students from various specialties, including humanities (history, philosophy, law, management), natural sciences (biology, ecology, chemistry), and exact sciences (physics, informatics, engineering). The first version of its textbook was published by TSU Publishing House in 2004 and the latest (updated and more extensive) in 2017 by a Moscow-based publisher "КноРус".

Now, an English translation of this book (slightly enhanced) is available. It sums up the applicable findings of many theoreticians and practitioners of systems thinking and design thinking (some of them are mentioned in the text and included in the list of references).

Acknowledgments

I am very grateful to my wife Larissa Petrovna and son Vladimir who helped me in preparing this book for publishing. Prof. David Gillespi has brushed up my English in the book, and deserves particular gratitude. The author expresses his sincere gratitude to Ms. Erin Harris, Senior Editorial Assistant, Taylor & Francis, whose recommendations to make certain changes and additions to the text has significantly improved the book, and Ms. Sofia Buono, Project Manager.

Author

F. P. Tarasenko, D.Sc. (Cybernetics & Information Theory), was born in Saratov, Russia on March 6, 1932. He is Chair Professor of Theoretical Cybernetics at Tomsk State University. He graduated from Radio-Physics Department of TSU in 1955. Since then, his career developed at TSU: lecturer at Chair of Radio-Physics (1958–1960); docent and head of the Chair of Electronic Computers and Automation (1960–1964); head of Chair of Statistical Radio-Physics and Information Theory (1964–1965); senior research worker in the laboratory of electronic computers (1965–1970); senior lecturer in physics, as a UNESCO expert in Dar-es-Salaam University College, Tanzania (1967–1968); head of Cybernetics Department of Siberian Physical-Technical Research Institute at TSU (1970–1977); and head of Chair of Theoretical Cybernetics (1977–1998). Since then a professor and Dean of International Department of Public and Business Administration, TSU (1992–2012).

He is the author of 199 scientific articles (10 monographs among them). He is also an Honored Worker of Science and Technology of Russian Federation (1990), Corresponding Member of Russian Academy of Natural Sciences (1991), and Full Member of International Academy of Sciences of Higher School.

Introduction: How Appeared the Systems Analysis

Perhaps everyone will agree that any human activity, regardless of its professional nature, aims to solve *problems* constantly and consistently arising before us. The problems are small- and large-scale, relatively easy and difficult, very different, and requiring the use of scientific and practical information and knowledge of various fields.

It is striking that some people more often successfully solve problems, while others experience difficulties and even failures. The natural desire for success prompts us to find out how the actions of successful and unsuccessful people differ. There is a need to accumulate and generalize the experience of solving problems, both positive and negative, in order to not repeat the wrong actions and use successful techniques in the future.

There are always specialists who seek to meet the public need: the study of the experience of problem-solving has also begun. However, what happened could be called a "historical misunderstanding". To solve any problem, it is necessary to use knowledge, often deeply professional, and a set of necessary professions for each problem is both specific and unique. This created a lasting impression that although everyone has problems, the problems of a doctor are very different from the problems of an engineer; the problems of a natural scientist are not the same as the problems of a manager; and so on. The specifics of the problems came to the fore. Therefore, the accumulation and generalization of experience in problem-solving began within each profession separately. In each specialty, such sections appeared, and in most professions, they took shape as entire disciplines. First, the military and then economists came up with "operations research"; physicians with "general human pathology" and "the art of diagnosis"; engineers with "system engineering" and "methods of engineering creativity"; social scientists with "political science", "futurology", and "conflict-ology"; and administrators with "systems approach" and "governance"; this list is not exhaustive.

In the 50s–60s of the last century, in the wake of the boom of cybernetics and systems sciences (now it is not possible to determine who first said "a"), appeared the idea to compare problem-solving methods in different professions. At first glance it appeared strange, but on a second glance, it appeared to be a natural phenomenon. Yes, to solve a specific problem you need special, sometimes very deep professional knowledge, specific to the problem. But if you pay attention not to the substantive specificity of the problem but to the technology of tackling with it, and to the sequence of actions and precautions, it turns out that the probability of success increases if you follow the same advice, regardless of the nature of the problem.

From this arose the idea of proposing a certain universal algorithm of actions for solving problems, suitable for use in any profession. This idea does not seem fantastic if we take into account that we all live in the same world, obey the same general laws of the universe, and only interact with it from different sides. This universal systemic approach was gradually realized by all, although some professionals still put a distinct professional meaning in their term "system analysis", describing the problems of their specialty.

For several decades, the idea of developing a commonly used problem-solving methodology led to the creation of a special technology, which we began to call ***Applied Systems Analysis*** (as opposed to concrete "system analyses"). This area of knowledge has already become a profession: systems analysts are trained in a number of universities around the world; there are dozens of companies taking orders to solve any problem from any customer; in Vienna, Austria, the International Institute for Applied Systems Analysis is working on global and international problems; and many higher education institutions include a course of applied systems analysis in the curricula of various faculties, both physical and mathematical, as well as natural sciences and humanities.

The technology of applied systems analysis can be compared with a locksmith's suitcase containing a set of tools and accessories that the locksmith uses when fixing the latest problem. In addition to the tools themselves, the locksmith uses the knowledge that must be applied in a certain sequence. Similarly, to use the technology of applied systems analysis, it is necessary to understand and accept its methodology, as well as to acquire a systemic view of the surrounding reality. Therefore, the entire course consists of two parts: (1) systems thinking: the methodology (philosophy and theory) of applied systems analysis; and (2) systems practice (design thinking): the technology of applied system analysis.

In the first part of this book, four basic concepts are presented on which the entire structure of this discipline is based: (1) the concept of a problem (how we evaluate the perceived reality), (2) the concept of a system (how reality is arranged), (3) the concept of a model (how we cognize reality), and (4) the concept of governance or control (how we change reality). They are sufficient for a logical, reasonable presentation and a conscious use of systems analysis technology, which is presented in the second part.

An important feature of applied systems analysis is to take into account the differences between the problems intentionally formalized, "hard" (up to the construction of quantitative mathematical models) and poorly structured, "soft", qualitative problems, set out in terms of spoken or descriptive professional language. Accordingly, different "hard" and "soft" methodologies are used in systems analysis. At the same time, the methods of gradual development, promoting our description of the problem from its "soft" appearance to the most "hard" option available in the given conditions, are developed.

Applied systems analysis differs from other sciences in a number of peculiar features.

First, it focuses not on finding general laws of nature, but on tackling a particular problem with its unique peculiarities.

Second, solving a real-life problem may require knowledge from several different professions; hence, applied systems analysis has a universal, interdisciplinary, over and above disciplinary character.

Third, the debate about the extent to which applied systems analysis can be considered as a science has ended with the understanding that a fusion of science, art, and craft is needed to solve real-life problems. The proportions between them are specific to each problem.

Fourth, systems analysis is performed not by the system analyst, but by the participants of the problem situation. The analyst knows the technology, that is, which questions and in what order to ask, and the answers to them are known only to the people involved in the situation. Hence, the product of the systems analysis (a decision to the problem) is developed not by a professional specialist (facilitator), but by a team of participants of the situation (stakeholders) under the unobtrusive guidance of the analyst.

Questions and Tasks

1. Why has the accumulation and generalization of experience in solving problems started (and continues) within each individual profession?
2. Why, despite the huge variety of problems, the technology (set of techniques) to solve them is almost the same in case of success and differs in case of failure?
3. Can you formulate the main differences between the applied systems analysis of traditional sciences?
4. Why can applied systems analysis be called a supradisciplinary and interdisciplinary field of activity, both in theoretical and in practical spheres?

Part I

Systems Thinking: Four Basic Concepts of Applied Systems Analysis

1 The Problem and Methods of Its Solution

Before discussing ways to solve problems, it is necessary to define the very concept of a problem. It is based on the original concept of a problem situation.

A problem situation *is a real set of circumstances, a state of things, that someone is unhappy with, dissatisfied with, and would like to change.*

This definition is illustrated in Figure 1.1. Now we concretize the concept of the problem.

The problem *is the subjective negative attitude of the person to reality.*

Let us pay attention to three points.

First, our definition fits any problem, regardless of its origin. Thus, we began to fulfill the promise to build a universal method of dealing with problems.

Second, in terms of the problem and the problem situation, two aspects are inextricably linked: objective (the presence of a real situation) and subjective (a negative assessment of reality by the subject). The difference between these concepts lies in what the emphasis is on: the "problem situation" highlights the objective component (reality), and the "problem" highlights the subjective one (dissatisfaction).

Third, there are no problems around us: the problem is a special state of the subject's psyche.

What does "solve the problem" mean? According to the definition, it is clear that anything should be done for this if only *to reduce or completely remove the discontent of the subject.* In the future, such a subject will be called a "client", and the person helping in solving the subject's problem will be called "systems analyst" or "facilitator".

1.1 PROBLEM-SOLVING OPTIONS

There are a number of ways to solve problems. Which one or what of them to apply in a particular case is decided by those who are engaged in solving the problem. But now let us discuss the possible options.

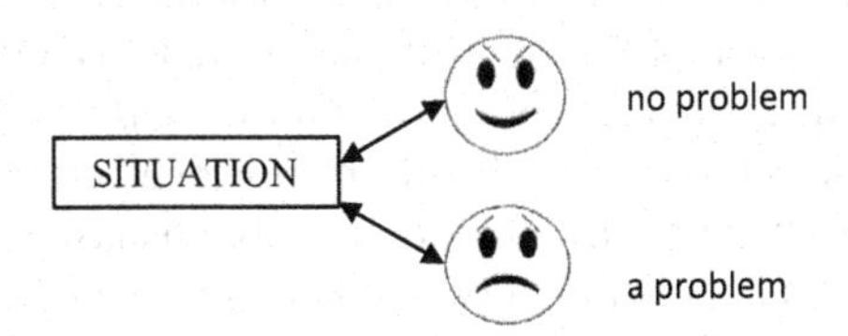

FIGURE 1.1 Work on performance review — need to add acknowledgments, do global role profiles.

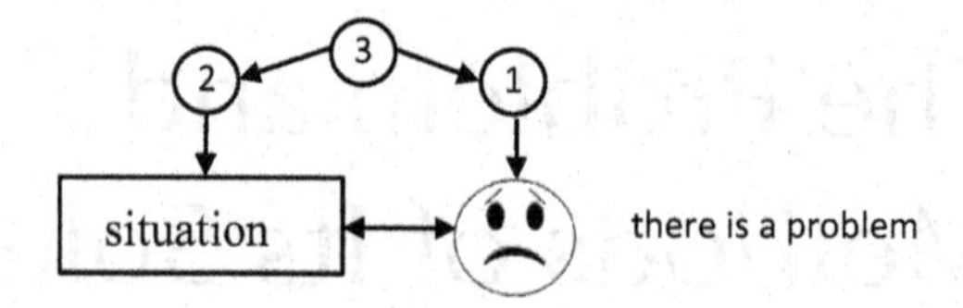

FIGURE 1.2 Reason of a problem may lie either in the subject, object, or both.

They are naturally divided into three groups: (1) to influence the subject to reduce his/her dissatisfaction, without changing the reality; (2) to change the reality so that the dissatisfaction of the subject is weakened; and (3) to arrange a proper combination of both possibilities (see Figure 1.2). Let us consider each of the groups.

1.2 WAYS TO INFLUENCE THE SUBJECT

There are three possibilities to change the attitude of the subject with respect to reality for the better, without changing the reality itself.

First, what is the subject dissatisfied with? This is it with what he knows about the situation. But he does not know everything! Among the things that he does not know, information of a positive nature could be one. If you inform the subject about this, his dissatisfaction will decrease.

While there are many examples of this, the one case that deserves special attention is when this is carried out in the form of education and training of the subject. In this case, the cause of dissatisfaction is precisely the lack of information, and getting the required information during training leads to a solution to the problem. Interestingly, when familiarizing with several American firms practicing systems analysis, it was found that about 80% of the problems of their clients were solved through training, retraining, and advanced training of the client's company personnel. This illustrates the fact that if you want to change reality, change yourself first.

It is worth noting another peculiarity of this problem-solving method. Additional information provided to the client must necessarily be positive, but not obligatorily true. There are cases when the problem is resolved with the help of false information. Everyone can remember an episode from his/her life when they were telling a lie. If you admit to yourself why the deception was preferred to truth, it turns out that with the help of lies in those conditions, it was possible to reach the goal much faster and easier than with the help of truth. This is not an excuse, and certainly not propaganda of a lie, but only a statement of the fact that there would be no lie if it were not useful. In all languages, there are concepts analogous to the Russian "lie for salvation": "white lie" or "holy lie" in social life; disinformation of an enemy in war; fake news in politics; mimicry among animals, insects, and even plants; and so on.

Another option for manipulating information is sorting out the useful truth from the harmful one, or preparing filtered half-truths. For example, one Dutch poultry farm managed to significantly increase the productivity of meat production by setting chickens' eye lenses with a darkened top. Among chickens there is a hierarchy: the larger the bird's crest, the higher it is in the hierarchy. During feeding, "seniors" drive away "juniors" from the trough. As lenses do not allow a bird to see who has what crest, disputes ceased, the food being stopped "dosed", and the growth of all birds increased dramatically (by about 20%) (See Figure 1.3).

The next possibility to solve the problem without changing the reality is to change the subject's perception of the reality. Since the evaluation of the relationship of a subject with the environment is a mental phenomenon, there is the possibility of influencing the psyche of the subject in the right direction. Forms of influence can be different: mental (hypnosis, suggestion, propaganda, advertising, etc.), physical (effects of various fields, such as acoustic, electric, or magnetic), and chemical (psychotropic drugs, narcotics, alcohol) (see Figure 1.4).

Let us emphasize that we do not evaluate what is good and what is bad; we merely state that there are actual opportunities (which must be used cautiously).

The third possibility to solve the problem without changing the problem situation itself is based on the fact that the problem arose as a result of the interaction of the subject with the situation. Therefore, sometimes the problem can be solved by interrupting this interaction (see Figure 1.5).

Here, too, there is a whole range of options: from pleasant ones to the problem carrier (promotion, assignment to study, or vacation), using more or less neutral (transfer to another department, rotation), to painful ones (dismissal, etc.), and even to the extremely cruel, condemned, but, unfortunately, still existing ("There is a person — there is a problem, there is no person — there is no problem").

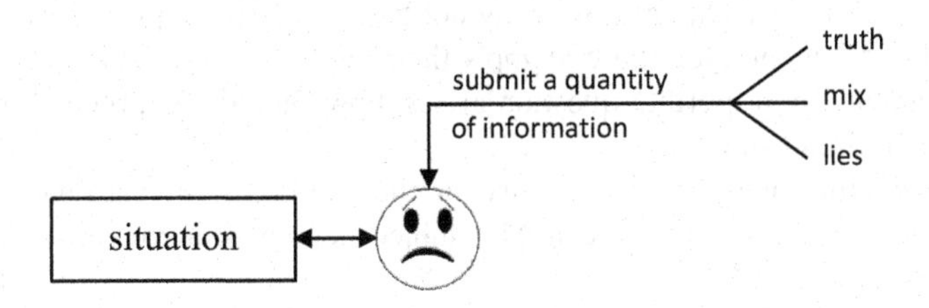

FIGURE 1.3 Some problems stem from shortage of information about the situation.

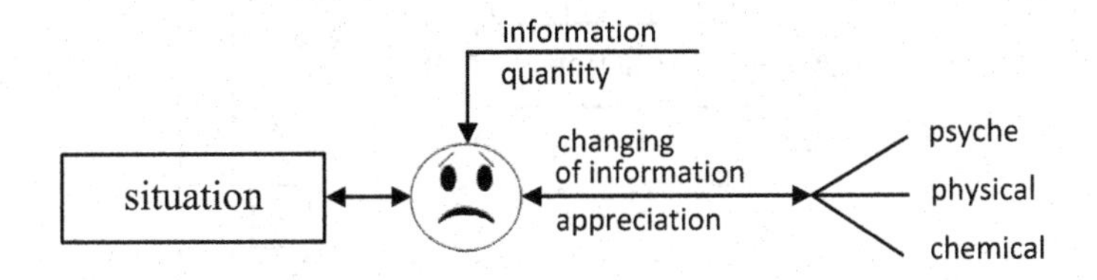

FIGURE 1.4 Some problems stem from an incorrect estimation of the situation.

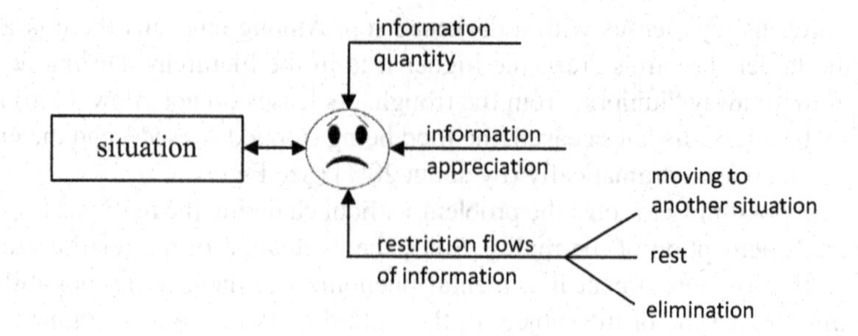

FIGURE 1.5 Some problems are inherent to the interaction between the subject and the situation.

1.3 INTERVENTION IN REALITY

Let us now turn to the second group of possibilities for solving the problem — by intervening in the problem situation itself. Naturally, the intervention should change the situation in such a way that the client's discontent decreases or disappears altogether. However, at the same time, we have to face a very significant circumstance, which, in fact, gave impetus to the detailed development of the technology of applied systems analysis. The fact is that in a real (problematic for our client) situation, not only our problem-holder is involved but also many other actors who assess this situation from their own positions. For them, it may not be a problem, or their problems may differ from the client's problem (see Figure 1.6).

Any change in the situation as a result of any intervention will be noticed and evaluated by all its participants, and may not be necessarily approved by all. Those displeased with the intervention will apply their resources to resist it.

A fundamentally important question arises: how should one proceed in connection with this circumstance?

To answer this question, let us turn to the fundamental, cardinal difference between the object and the subject. The subject, being simultaneously a physical

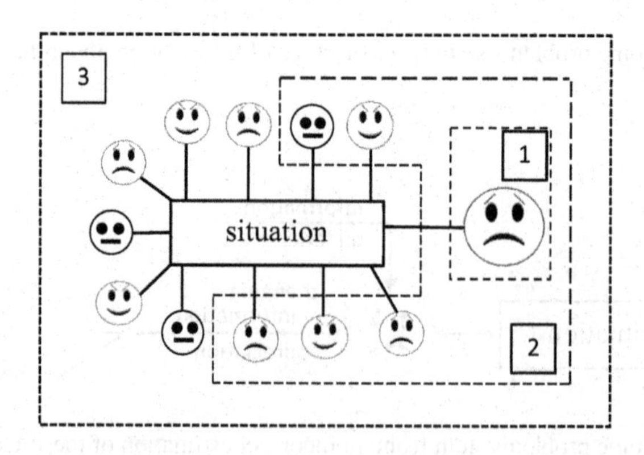

FIGURE 1.6 Three options of treating problem according to different ideologies.

object, exists in a real physical environment and, like any other object, is subject to the effects of this environment. Unlike the object, the subject is not only subordinate to natural laws but also endowed with the ability to evaluate its interactions with the environment: he may or may not like something. This is where the individuality of the subject is laid. Subsequently (in the chapter on models), we will discuss the reasons for this, but for now, we emphasize that the assessments are purely individual and subjective and that there can be no objective assessments. As a result, the same reality is evaluated by different subjects differently.

The following advice may be useful in this regard:

> *Whenever any evaluative word is heard in your presence* (good–bad, useful–harmful, right–wrong, etc.), *be alert, and ask the question:*
> **"In what sense?"**

The essence of the advice is that no evaluations are objective. Evaluations are always subjective, and if you want to understand the true meaning of what has been said, you need to find out what criteria the evaluator applies as different subjects may evaluate the same thing differently.

Let us now return to our question of how to act, solving the client's problem if there are other participants in the situation with inevitably different interests. Answer: we must act correctly. The word "correct" is an evaluative one; hence, the question arises of what is meant by this.

1.4 THREE TYPES OF IDEOLOGIES

The *correct* behavior is considered to be the one that is most consistent with the ideology adopted by the subject. It is the ideology that determines what is bad and what is good, what is right and what is wrong.

It turns out that ideologies may be different. The adherence to one's "own" ideology is a complex result of personal choice based on the impact of education, culture, and circumstances. Ideologists cite a large number of arguments in favor of their own ideology, discussing its many differences from other teachings. However, you can point out one feature that helps to distinguish between ideologies in our case. This is the definition of *what attitude to other subjects is correct.*

Although many various gradations between ideologies can be introduced (like between numerous political parties in some countries), the essential differences in attitude to others can be made between the three types of ideologies. Each of these ideologies leads to different approaches to solving the real-life problem.

The first type of ideology is called conditionally "*the principle of the priority of the major person*". In our case (Figure 1.6, field 1), this principle leads to an intervention that is pleasing to the client, irrespective of the opinions of other participants. Some of them may like it, and some may not, but it must be implemented by all means. There are real-life examples of implementation of such ideologies: dictatorship, monarchy, unity of command, hierarchical organization, egoism, self-esteem, etc. You can even explicitly admit that in some circumstances, such an ideology gives the greatest chance for success (army, war, emergency, need to concentrate efforts of many people on a single common purpose, etc.).

It is necessary, however, to keep in mind that a number of specific features are inherent in the implementation of this ideology, which will inevitably have to be taken into account. First, the implementation of such an approach to solving the problem of the number one person will surely cause discontent of a certain part of the other participants in the situation, which will prompt them to respond. Hence, those who accepted this ideology should have the power to suppress the discontented and the willingness to use force.

The second ideology can be called the "*group priority principle*". According to it, among the participants of the situation, besides the client, there are other subjects no less important and valuable than the client (Figure 1.6, field 2).

Therefore, now the intervention should be carried out taking into account the interests of all "ours". This, on the one hand, complicates the design of the intervention, but on the other, it opens up the possibility of using the resources of not only the client but also the rest of the group. There are many examples of real practice of this ideology: racism, nationalism, fascism, communism, or any group activity, including political party, trade union, sports, etc. We emphasize that here we do not set the task to assess such ideology: for those who accepted it, it is the only correct one, while for opponents, it is unacceptable. However, it is worth noting some of the inherent features of this ideology, which are latently embedded in it and in appropriate conditions may manifest themselves negatively. First, this is a double morality: by dividing everyone into "us" and "them", "ours" and "alien", it allows them to be treated differently. In the class variant, "not ours" are generally regarded as enemies, which leads to aggressiveness and build-up of power structures. Different groups resolve this contradiction in different ways, and stories are known for successful and tragic variants. The case when a "major" person belongs to several different groups simultaneously deserves special attention, which is an inherent reason for corruption.

And now, the good news: applied systems analysis adheres to the third ideology (see Figure 1.6, field 3), the "*the principle of priority of all and each*", which is based on two postulates:

- there are no identical subjects, and they are all different;
- despite the differences, all subjects possess equal rights and have equal social value.

It follows that it is wrong and immoral to solve the problem of some at the expense of others. Only improving intervention is the recognized correct and moral approach.

Improving intervention *is a change in a problem situation that is positively evaluated by at least one of its participants and non-negatively by all others.*

Naturally, our client should be among those who positively assess the proposed intervention.

In connection with the above, applied systems analysis can be called *the theory and practice of design and implementation of improving interventions.* Since this does not give rise to new discontent among any of the participants in the situation, another (equivalent) definition of applied systems analysis can be formulated as *a technique for solving real-life problems without creating new problems.*

1.5 IS IMPROVING INTERVENTION FEASIBLE?

The experience of reading this course shows that when the definition of improving interventions is announced in the audience, there necessarily found skeptics (and the older the audience, the more of them), who consider what has been said to be a beautiful but unattainable goal. There are many reasons for skepticism. Let's discuss the main ones.

- ***"It is impossible to make everyone feel good".*** This is a substitution of the thesis: improving intervention is not when "everyone is good", but when no one is worse, and this is not the same thing.
- ***"Experience shows that no solution to any complex problem is possible without creating new problems".*** In reality, it often happens. But this is only a consequence of the fact that the systems technology was not realized properly. Obvious examples of this are the antialcohol campaigns conducted in United States in the beginning of the 20th century, and in the end of it in Russia during the Soviet era. Both attempts failed and gave rise to many new problems. The negative experience of solving problems is not connected with the impossibility of solving them successfully, but with noncompliance with the requirements of their systemic solution.
- ***"Very often failures in solving problems are due to lack of resources and mistakes in decision-making".*** Therefore, it is important to consider improving intervention as an ideal to be pursued, even if it turns out to be not fully achievable. "Improving intervention is often difficult to find, but rarely impossible" (R. L. Ackoff [1]). Applied systems analysis proposes a method of moving toward a goal, although in practice this may be hindered by a lack of information, making mistakes, insufficient resources, and a shortage of time. It is important to move in the right direction as far as possible (this will be discussed in detail in Part II of this book).

Perhaps the most serious reason for doubting the feasibility of improving interventions is the contradictory interests of the participants in the problem situation, sometimes reaching the point of conflict. How can you make an improvement intervention when someone is dissatisfied only with the fact that some others are well? Or when everyone seeks to prove and assert their rightness, and their positions are different or even incompatible?

There are several possibilities to move toward the improved intervention even in such conditions (see below). In fact, in systems analysis, it is offered to refuse clarification "who is right?", and move to the search or creation of "common agreement", preferring common peace and efficiency ("consensus") before their rightness.

1.6 FOUR TYPES OF IMPROVING INTERVENTIONS

Returning to the ways of solving problems, it is interesting to consider the classification proposed by R. Ackoff [1]. He noted that despite the enormous diversity and the dissimilarity of problems, there are only four ways to solve them.

1 **ABSOLUTION**

This term in colloquial English denotes the actions of a priest who forgives sins to parishioners: he listens to confession and *does nothing*. In the professional language of systems analysis, this term denotes noninterference. Please note that this does not make anyone worse. However, preference should be given to noninterference only if any proposed interventions lead to worse results. For example, actions of a doctor in case of a difficult diagnosis (placebo), behavior of a sapper when an explosive device is unknown to him, and recommended non-interference in the family problems of your friends or spouses.

2 **RESOLUTION**

In such a type of intervention, the problem is solved partially, not fully, but in an acceptable manner. There are several possibilities to do this.

The first is to use insufficient resources for completely solving the problem to mitigate discontent in having solved the problem partially. A good example of this is some increase in wages, pensions, and scholarships to public-sector employees against the background of galloping inflation or allocation of limited resources by draw, by turns, or equally.

The second possibility is to try to return to the state when there was no problem: determine what caused the problem and eliminate the cause, that is, find the culprit and punish him. This (alas common) solution is incomplete, partial, and outdated: any event in the world is the result of many factors, and the elimination of one is certainly not adequate.

Another example of such an approach is to repeat the action that was previously successful in a similar case. But this involves the risk of insufficient similarity of circumstances, which can lead to unexpected results.

3 **SOLUTION**

In the professional language of systems analysis, this is the term denoting the best intervention under given conditions. The relevant scientific term "optimal" has already entered the spoken language and public consciousness, so it is important to understand and apply it correctly. *Optimal means the best under the given constraints.* For all the seeming simplicity of this definition, it requires explanation.

First, what does "the best" mean? The same objects can be ordered in different ways, depending on what quality to consider. The criterion that evaluates this quality allows you to find the best (for this quality) alternative. How to choose the best option if the alternatives are compared not by one but by a combination of several criteria? This is not a trivial question. Let us consider (as it will turn out later, erroneously) this question as purely technical and consider it in the "Choice" section of the second part of the course. Meanwhile, it's important for us that before talking about optimality, it is necessary to specify and determine by which criterion (or criteria) the compared options will be ordered, that is, in what sense we will use the term "best".

However, this is not enough for optimality. The second, no less important, integral part of the concept of optimality is the dependence of the result of choice on the specific constraints in this situation. Under the same quality criteria, the choice from the same set of alternatives under different constraints will generally be different. Therefore, only those alternatives that satisfy the imposed restrictions should be compared with each other according to the chosen quality criterion: the best alternative in the sense of the criterion that does not meet the restriction cannot be implemented.

The desire to do everything as best as possible is so natural for people that it is not surprising how quickly the abstract mathematical concept of optimality has moved from science into business, governance, and even in everyday life. Although the widespread popularity of the idea of optimality (apparently, it was a consequence of the great fashion for cybernetics in the 1950s–1980s), few people, except specialists, drew attention to the warning of N. Wiener [2], the father of cybernetics, about the need for a careful use of this concept.

The fact is that intervention in a problematic situation is based on the information we have about the situation and the degree of knowledge regarding our situation may be different. There are well-structured problems that allow the construction of quantitative mathematical models, for example, many engineering and scientific problems (such problems are called "hard" problems). But many real-life problems are described in languages far from mathematics, from everyday problems described in spoken language, to professional ones, for example, many humanitarians and naturalists (such problems are called "fuzzy" or "soft" problems). Naturally, these differences are also taken into account in a number of details of the problem-solving technology, which makes it possible to speak of "hard" and "soft" technologies of applied systems analysis.

The difference between "hard" and "soft" methodologies of systems analysis can be emphasized once again considering optimality. Both optimality components (criteria and constraints) are sensitive to this difference. The very requirement that the quality criteria and limitations are expressed quantitatively implies that the situation under consideration is so well-studied that it allows the construction of its mathematical description. The optimization problem is a formal mathematical problem, quite adequate to the problems of the "hard" type (and the course of optimization methods is one of the largest and most ingenious mathematical subjects at the university).

However, within the framework of formal mathematical models, the "fragility" of optimal solutions was revealed; often even with small deviations from the assumptions in the formulation of the problem, the quality of its solution can change very dramatically. Therefore, consideration of the problems of stability (robustness) of solutions is an important section of the theory of optimization.

In the transition to "soft" problems, the situation is much more complicated. It's not just that for such problems that it's harder (if at all possible) to find quantitative measures for criteria and constraints. The main thing is that the "softness" of the problem is a consequence of little knowledge about the problem; in particular, there is no possibility to list all the important limitations, which, as we have seen, radically affects the quality of choice. Hence, the optimality in this case should be considered an unattainable ideal, which is still worth striving for. Additionally, optimization attempts should be considered only as an element of the trial and error method discussed at the end of this part.

4 **DISSOLUTION**

Dissolution denotes an intervention that ends in complete extinction of the problem and nonappearance of new problems. It would seem what could be better than optimal? The essential difference between the third and fourth methods is that "optimal" is the best under given conditions, and "dissolution" considers the restrictions and conditions not as firmly fixed, but as subject to change or cancellation to find new and previously unacceptable options, among which may be options that are much more effective than previously optimal.

An important option to "dissolve" the problem is to prevent it by taking measures to ensure that it does not appear. Here, the change of the system is made not *after* the appearance of the problem to solve it but *before* that to prevent it.

A visual analogy is the change in the fire protection system. In the past, this system mainly consisted of firefighters, brave, exotically dressed, and rushing down the street in their cars with flashing lights and sirens to fight fire and heroically save the lives of people and their property. In the current system, the majority of employees are fire inspectors. They regularly check the buildings under their jurisdiction without light and sound effects to prevent fires. They prescribe improving changes in objects and processes and reducing the likelihood of fire and possible damage from them. This activity is no less — if not more! — important than extinguishing fires.

In this way, in far-sighted organizations, special employees are introduced into the staff, whose duty is to conduct preventive inspection of all components of the organization, as well as early detection of dangerous situations and trends. Sometimes such employees are called "internal auditors", "thinking engineers", "inspectors", etc. So far, this function is implemented only in the form of an internal audit. A gap in organizations is the absence of persons responsible for implementing preventive improving intervention. This responsibility is assigned to managers who are already loaded with the management of the current activities of their unit.

R. Ackoff gave a vivid example of applying all four methods to solving one real problem.

A problem has arisen in a bus company in a large city: after the introduction of performance premiums for high quality, conflicts between drivers and conductors began. The quality of the drivers' work was assessed by the accuracy of the observance of the timetable, and that of the conductors' by how well they serve the passengers. During rush hours, the conductors delayed the sending signal (they had to check not only the ticket availability but also the correct payment depending on the distance, and the incoming ones to sell the tickets, each to his destination station), and this adversely affected the drivers' surcharge.

At first, the company's management ignored the problem (ABSOLUTION), expecting that everything would settle down by itself. But the problem continued to escalate, and the trade unions were involved in the conflict. Subsequently, the leadership tried to return to the old system of payment (RESOLUTION), but both trade unions protested as this would mean the abolition of surcharges. The management then suggested to the trade unions to agree on the division of the surcharge fund (SOLUTION), but they did not want to cooperate with each other.

The problem was DISSOLUTED by a visiting systems analyst who discovered that during rush hours the number of buses on the line (regulated during the day) was greater than the number of stops. During these hours, the conductors began to be removed from the buses and instead were assigned to each stops. They started selling tickets before the bus arrived, had the time to check the tickets for those leaving, and began to send a departure signal on time. At the end of the rush hour, the conductors returned to the buses, and the extra buses were removed from the line. In addition, the company hired a smaller number of conductors.

This interesting example should not give the impression that the four types of solutions are arranged in order of absolute preference: in this case, "dissolving" (removal the restriction "conductors must always work in the bus") turned out to be the best of them, but in other problems any other option can turn out to be the best.

Moreover, the preference for a particular type of intervention depends on the mental orientation of the manager. R. Akoff proposed [1] to distinguish four types of managers engaged in planning, decision-making, and implementing decisions:

1. ***Reactive management*** is dissatisfied with the current situation and where everything is going; it prefers what was in the past; and its efforts are aimed at returning to the previous state by eliminating the causes of the changes. The preferred type of problem-solving is *resolution*, and the methods used are past experience, common sense, qualitative assessments, the choice of a "good enough", and "acceptable" solution. An example of a satisfactory application of such an approach is the clinical practice of healing, but even then there are fatal failures. In management, such an approach is associated with authoritarian management, planning from the top down, aimed at solving separate problems and eliminating the undesirable without considering its connection with the other components of the situation (which often leads to even more undesirable problems).
2. ***Inactive (passive) management*** is satisfied with the present and wants neither a return to the past nor future changes; impedes changes and appreciates stability; believes that if nothing is done, then nothing will happen, and that is good; believes that it is necessary to act only when there is a threat or a crisis. At the same time, unlike reactivists who are trying to eliminate the causes, inactivists are engaged in the suppression of symptoms ("crisis management"). The preferred type of problem-solving is "*absolution*", ignoring or denying the problem, and the hope that it will disappear or be resolved by itself.

3. ***Proactive (preventive) management*** is convinced that the future will be better than the past and the present; therefore, it tries to accelerate changes and use the opportunities associated with them. Forecasting the future, the ability to learn and adapt to changes in the environment, planning, and creating changes becomes important. The preferred type of problem-solving is "*solution*", finding the optimal solution, that is, the best in given conditions. Technologies are mainly quantitative, such as methods of optimization, operations research, mathematical (more often, linear) programming, risk analysis, balance of expenses and income, etc.
4. ***Interactive management*** not only does not want to return to the past and the perception of the present but also to accept the impending future. It is sure that the future can be created by the efforts aimed at it. The preferred type of solution is *dissolution*, the implementation of changes in the system and/or its environment, leading to the disappearance of the problem. The technology of this is *idealized design.*

1.7 MORE ABOUT APPLIED SYSTEMS ANALYSIS

This chapter has two objectives: to specify the concept of the problem and methods of solving it, and, moreover, to give a general idea of the applied systems analysis itself. The first goal may be considered as achieved (to the extent we need now). To achieve the second goal, two more features of applied systems analysis should be discussed, which have not yet been mentioned.

Consider the typical sequence of actions in time during the systems analysis (Figure 1.7).

At the moment "Problem", the client turns to the system analyst with his problem, which he could not solve on his own. After signing a contract that imposes a number of obligations on both sides (which we will discuss later), work begins in accordance with the technology (described in the second part of the book). After a series of operations, there comes a moment, "Model", when we get a sufficiently adequate model (the exact meaning of these terms will be given later) of the problem situation. Now comes the period of using the model to obtain the results of certain interventions. At the end of this period, a (usually multicriteria) selection of the most appropriate option "Decision" is made. From the decision to its implementation, the path is not easy, and requires a fairly strict adherence to technology (in modern language, this is called management). With diligence and luck, we can reach the "End" point when

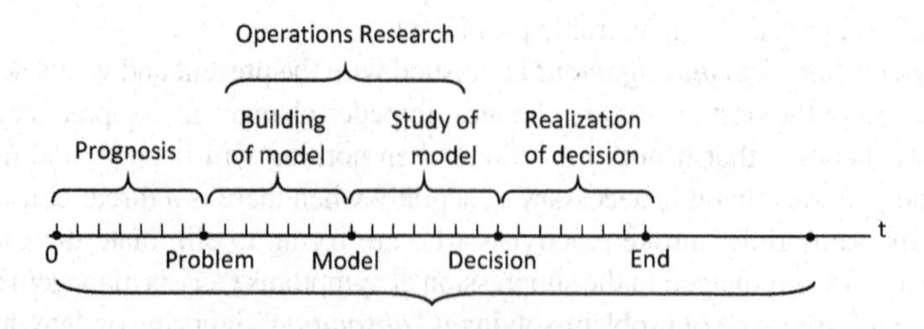

FIGURE 1.7 Scheme of the stages of systems analysis.

the problem is solved. A more detailed description of the operations involved in each stage, with constant care to maximize the probability of success in the presence of traps, the possibility of error, limited resources, lack of time, and incompleteness and inaccuracy of information will be the subject of the second part of the course. Meanwhile, let's pay attention to two more features of applied systems analysis.

The first follows from the fact that there was an "O" moment in the past (Figure 1.7) when there was no problem at all. If the client then turned to a systems analyst, one could subject analyzing the course of the future and predicting the appearance of a problem while maintaining the firm's style and tactics. But it would also be possible to design an intervention that would prevent the occurrence of the problem. This is reflected in a slightly humorous saying: "The best systems analysis is one that does not come true". Therefore, the forecasting technique is included in the arsenal of applied systems analysis. So far, this function is realized in organizations only in the form of internal audits. It is a fact, however, that clients most often turn to analysts after their own attempts fail to solve an already urgent problem.

The second very important and fundamental feature of applied systems analysis is indicated in Figure 1.7, coverage of the scope of analysis beyond the limit "End" of the problem. It allows you to discuss the question: what will happen after the problem is solved? Of course, the former client will again have some problem. Not as a result of the solution of the previous one if we tried to implement an improving intervention (in principle not creating new problems), but in case of unavoidable changes in the environment and in the system itself. Whether shall we go back to the consulting firm? This will not be necessary due to the specific feature of the systems analysis.

In fact, the problem-solving process cannot be performed only by the system analyst alone. To build a model of a problem situation, information possessed only by its participants themselves is needed. Therefore, a mandatory element of the technology is their involvement in the process of working on a problem. The system analyst knows what questions to ask, in what form, and in what sequence to ask, to build an adequate model of the situation on the basis of the information received, and only the participants in the situation can answer these questions. Moreover, it is they, and not the analyst, who will have to implement the developed intervention.

Consequently, the process of systems analysis will be performed by the employees of the client's company themselves. Doing some work on their own is the most effective form of training them for this activity. Thus, in applied systems analysis, it turns out to be a naturally built-in and integral part of training in systems analysis itself. As a result, the need to reapply to a consulting firm is significantly reduced.

Questions and Tasks

1. Explain the differences between the concepts of "problem situation" and "problem".
2. What does "to solve the problem" mean?
3. Which three ways to influence a subject without changing reality can (under certain conditions) lead to the solution of his problem? What are these conditions?
4. What is the main difference between a subject and an object?

5. How to determine the meaning of the assessment expressed by a certain subject?
6. Why do we have to rely on some ideology when intervening in reality to solve the problem?
7. Reproduce the classification of ideologies into three types. What is the main difference between them?
8. The goal of the applied systems analysis is to create an improvement intervention. List at least three reasons why this may not actually happen.
9. Name four types of improvement interventions.
10. Optimality is ensured only when two requirements are met together. What are these requirements?
11. What is the important result of the applied systems analysis of a specific problem, besides solving the problem itself?
12. This chapter introduces special concepts (and corresponding terms) that are included in the professional language of applied systems analysis. Some of these terms are used in the spoken language, but in a different, vaguer sense. Others have a special meaning. Check whether you can reproduce professional definitions for all the concepts listed below (if there is not one, be sure to find such a definition and try to remember it):
    - a problem situation;
    - evaluation (of something by the subject);
    - a problem;
    - solution to the problem;
    - intervention;
    - improving intervention;
    - applied system analysis;
    - optimality; and
    - "hard" and "soft" problems.

2 The Concept of the System

The term "system" is used in a very broad sense. We say solar system, Stanislavsky's system, nervous system, system of equations, heating system, system of views and beliefs, etc. Some systems naturally occur in nature, while some are artificially created by man (material and ideal). The language used reflects commonalities between all manifestations of reality. It is clear that the universe exists on a single simple principle: any distinguished entity in the world (a system, a thing, or a process) consists of some parts that are connected in a specific way and belong to larger system(s). It's amazing that this monotony engenders a huge variety of systems in nature. This universal similarity is denoted by the term "systemic", and we have to concretize its meaning.

The simplest description of the diversity of systems is their classification. But separating an itemized set of elements into a less number of different classes of "similar" elements may be done on various indications.

Some features of the system that have to be taken into account when working with it are associated with its origin. These features can be classified as follows:

- physical systems (e.g., river basins, galactic);
- biological systems (e.g., living organisms, populations);
- technical systems (e.g., cars, power plants, buildings);
- abstract systems (e.g., philosophy, mathematics, ideologies);
- social systems (e.g., family, ethnic group).

Other classifications are also possible, for example:

- ecological systems (e.g., co-existence of physical, biological, technical, and social systems);
- organizational systems (e.g., groups, states, political parties);
- process systems (e.g., algorithms, technologies, lifecycle).

R. Akoff proposed [3] to classify a system by how much inherent is the goal-setting and choice of parts of the system in the system itself.

Since the specific use of a term is related to a specific situation and the purpose of its use, there are many different definitions of the system, ranging from the most abstract ("system is everything we want to see as a system") to the highly specific definitions given by exact, natural, and humanitarian sciences (one philosopher has collected and published a collection of 45 different definitions!).

The specific definition of a system is required for systems analysis. By virtue of its focus on solving problems, the concept of a system in this case should be very general and thus applicable to any situation. The solution is to identify, list, and describe the features and properties of systems that, first, are inherent in all systems without exception, regardless of their artificial or natural origin, material or ideal incarnation. Second, from the multitude of various properties, we must select and include the obligatory ones that will be needed to build and use certain operation(s) in the technology of systems analysis. The resulting list of properties can be called a descriptive definition of the system.

Let's start discussing the necessary properties of the system, which naturally fall into three groups, namely, static, dynamic, and synthetic, each comprising four properties.

2.1 STATIC PROPERTIES OF THE SYSTEM

The static properties are the features of a particular state of the system. It is like something that can be seen on an instant photo of a system; in other words, something that a system possesses at any fixed moment of time.

Integrity is the first property of the system. Every system acts as single, whole, separate, or different from everything else. We will call this property *integrity of the system*. It allows the entire world to be divided into two parts: the system and its environment (Figure 2.1). Although the concept of integrity will continue to expand and deepen during further consideration of a system, for now it denotes only the external distinctiveness of the system in the environment.

Openness is the second property of the system. The system, which is distinguished from the rest, is not isolated from the environment. On the contrary, systems are connected and exchange among themselves resources, such as substance, energy, and information, in proper quantities and qualities of each. We denote this feature by the term "openness" of the system and discuss this property in more detail.

Note that the connections between the system and the environment are directional: through some, the environment affects the system (called *system inputs*), while through others, the system has an impact on the environment, does something in the environment, and gives something to the environment (such connections are called *system outputs*) (Figure 2.2).

The list of inputs and outputs of the system is called the *black box model*. In this model, there is no information about the internal peculiarities of the system. Despite the simplicity and poor content of the black box model, it is often quite sufficient to work with the system successfully. In many cases, processing on equipment (car, radio, computer, device) or on people (e.g., in management), only input and output information of the corresponding system helps to achieve the goal. However, to do this, the

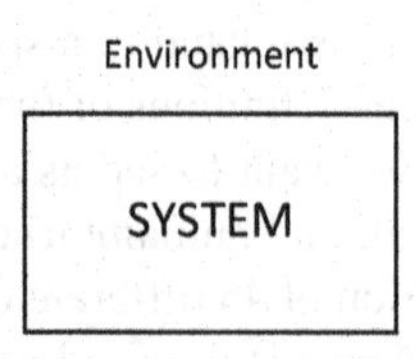

FIGURE 2.1 A system is a separate part of reality.

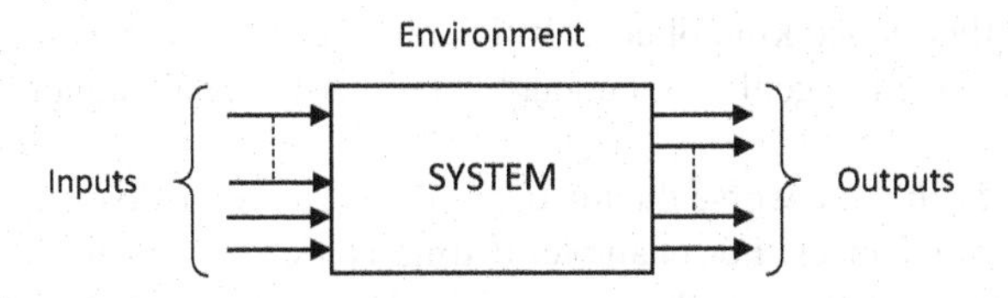

FIGURE 2.2 Input and output ties between the system and the environment.

model must meet certain requirements. You may experience difficulties if you are not aware of all the inputs of a system, such as if you do not know that in some models of TVs the power button should not be pressed, but pulled, or that in some hotels the switch in a dark room is combined with the door-side valve. It is clear that to successfully manage the system, the black box model must contain all the information necessary to achieve the goal. In attempting to meet this requirement, the decision-maker will encounter difficulties that should be borne in mind. Let's list these difficulties.

The difficulties of building a black box model. All of these difficulties stem from the fact that the model always contains a finite list of links, whereas their number in a real system is not limited. The question arises: which of them should be included in the model and which should not? We already know the answer: the model should reflect all the connections that are essential for achieving the goal. But the word "essential" is evaluative! Evaluation can only give a subject. But in addition to the ability to evaluate, the subjects need to have another property, the ability to sometimes make mistakes in their assessments. An error in the estimate will cause the model to not fully meet the requirement of adequacy, and its use will lead to difficulties in working with the system.

There are four types of errors when building a black box model.

An *error of the **first** kind* occurs when the subject regards the relationship as significant and decides to include it in the model, whereas in reality it is irrelevant in relation to the goal set and could be neglected. This results in redundant elements in the model, which are essentially unnecessary.

The ***second** kind of mistake*, on the contrary, is made by the subject when he decides that a relationship is insignificant and does not deserve to be included in the model, whereas in fact without it our goal cannot be achieved in full or even at all.

Monitoring question: which of the errors is worse? Usually they say, of course, the second. This answer is not accurate. After all, the word "worse" is an estimating one! Therefore, it is necessary to define "in what sense?". Note that using a model that contains an error will inevitably lead to losses. Losses can be small, acceptable or intolerable, or unacceptable.

If the quality of the solution is determined by the amount of losses in its implementation, the question of which error is worse is reduced to comparing the values of losses associated with them.

The damage caused by the error of the first kind is due to the information entered is superfluous. When working with such a model, you will have to spend additional resources on fixing and processing unnecessary information, for example, spending computer memory and processing time on it. This may not affect the quality of the solution, but the cost and timeliness are necessarily impacted.

The loss of the second kind of error is the damage from the fact that the information to fully achieve the goal is not enough, and that the goal cannot be achieved in full.

It is now clear that the worse the mistake, the more are the losses, which depends on specific circumstances. For example, if time is a critical factor, the error of the first kind becomes much more dangerous than the second; a decision made in time, even if not the best, is preferable to the optimal, but late, decision.

The ***third** kind of error* is caused by ignorance. To assess the significance of a certain connection, one must know that it exists in general. If it is not known, there is no question of whether or it is included in the model at all; in our models, only what we know is included. However, even though we are not aware of the existence of a certain tie, it does not cease to exist and manifest itself in reality. Hence, it all depends on how essential it is to achieve our goal. If it is insignificant, then even we will not notice its presence in reality and its absence in the model. If it is significant, we will experience the same difficulties as with the error of the second kind. The difference is that the error of the third kind is more difficult to correct, which requires obtaining new knowledge.

An *error of the **fourth** kind* can occur if the known and recognized significant connection is incorrectly assigned to the number of inputs or outputs. For example, a strong correlation between the yield of grain and egg production of hens can be interpreted as an input, which is known, and the output, which needs to be evaluated. But you can consider wearing a certain headdress as an input, because it was prevalent in England in the last century, the health of men wearing cylinders, which was much better than the health of wearing caps. How can we interpret the fact that prisoners attend Church more often than people on the outside? And the problem of symptoms and syndromes in medicine?

Thus, when building a black box model, you should be careful to not make any of the four mistakes.

Openness of systems and integrity of the world. The universal interconnection and interdependence in nature is very important for the systems analysis of the openness of systems. This law of dialectics, established in the intellectual and experimental torments of several generations, is quite a simple result of the openness of systems. Between any two systems that necessarily exist, both long and short chains of systems, the output of each system is linked to the input of the other. In this case, the forward and reverse chains are usually different where the concept of asymmetric cause-and-effect relationship arises.

In conclusion, considering the second property of systems offers little intellectual fun. Answer the question: are there closed (i.e., not connected to the environment) systems? Choose the correct answer from the three possible ones: (1) yes, there are; (2) no, do not exist; and (3) I do not know, and never will know. As an experiment that tests the existence or nonexistence of a closed system is impossible to design, this question is a matter of faith, not science; supporters of opposite statements ("yes" and "no") are not able to prove their case, no matter how confident they are in it.

The ***third** property of the system: **internal heterogeneity and the distinguishability of parts***. If you look inside the "black box", it turns out that the system is not homogeneous, not monolithic: you can find different qualities in different places.

The description of the internal heterogeneity of the system is reduced to the separation of relatively homogeneous areas, drawing the boundaries between them. This is how the concept of the parts of the system appears. On a closer look, it turns out that the selected large parts are also not homogeneous, which sometimes requires to allocate even smaller parts. As a result, a hierarchical list of system parts is obtained, which is called the *system composition model* (see Figure 2.3).

Information about the composition of the system may be needed to work with the system. The objectives of interaction with systems may be different, and therefore the composition models of the same system may differ. It is not easy to create a useful, workable model.

The difficulties of constructing a model of composition. At first glance, parts of the system are easy to distinguish as they "catch the eye". Some systems can be differentiated into parts spontaneously in the process of natural growth and development (organisms, societies, planetary systems, molecules, mineral deposits, etc.). Artificial systems are deliberately assembled from previously separate parts (mechanisms, buildings, texts, melodies, etc.). There are also mixed types of systems (nature reservation territories, agricultural systems, environmental and ecological organizations, animal-drawn transport).

On the other hand, on asking what constitutes a university, the rector, the student, the accountant, and the infrastructure executive each will give its own model composition, differing from the others. Also, differently determine the composition of the aircraft pilot, stewardess, and passenger. Similarly, we can also say that our body consists of the right and left halves, as well as top and bottom halves. So, what does it really consist of?

The difficulties of building a model of composition, which everyone has to overcome, can be represented in three formulations.

The first. The whole can be divided into parts in different ways (as cutting a loaf of bread into chunks of different sizes and shapes). But how exactly? The answer lies in the way you need to achieve your goal. For example, the composition of cars in different ways represents novice motorists, future professional drivers, locksmiths preparing to work in car repair shops, and sellers in car shops.

Then it is natural to return to the question: are there parts "in fact"? Note the careful formulation of the property in question: the *distinguishability* of parts, and the *inseparability* into parts. Now, we again come to the problem of *system integrity*:

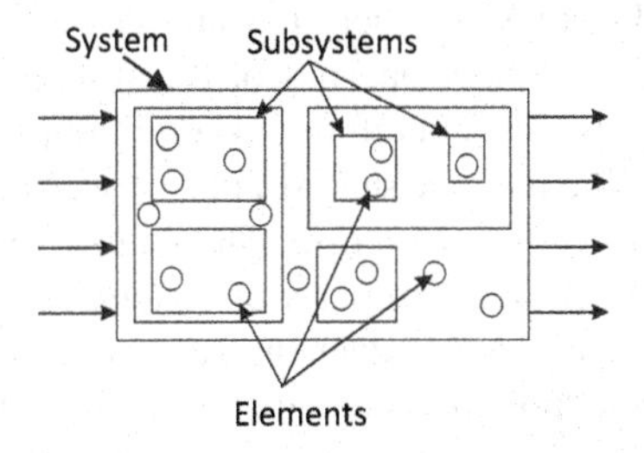

FIGURE 2.3 Model of the components of a system.

you can distinguish between the parts of the system you need for your purpose and use the information available to you about them, but you should not separate them. Later, we will develop this recommendation.

The second. The number of parts in the composition model also depends on the level at which the fragmentation of the system is stopped. The final parts on the branches of the resulting hierarchical tree are called *elements*. In various circumstances, the decomposition is terminated at different levels. For example, when describing upcoming work, it is necessary to give an experienced worker and a beginner instructions at varying degrees of detail. Thus, the model composition depends on what is considered elementary, and since this is an evaluative word, it is not an absolute but a relative concept. However, there are cases when an element is natural, absolute (a cell is the simplest element of a living organism; an individual is the last element of society) or is determined by our capabilities (e.g., we can assume that an electron also consists of something, but so far physicists have not been able to detect its parts with a fractional charge).

The third. Any system is part of a larger system (and often a part of several larger systems at once). This metasystem can also be divided into subsystems in different ways. This means that the external boundary of the system has a relative, conditional character. Even the "obvious" border of the system (human skin, enterprise fencing, shores of island, etc.) under certain conditions turns out to be insufficient for determining the boundary under these conditions. For example, during a meal I take a cutlet with a fork from a plate, bite it off, chew it, swallow it, and digest it. Where is the border crossing that cutlet becomes my part? Another example is the enterprise boundary. A worker fell on the stairs and broke his leg. After treatment, when a bulletin is paid for, the question arises: what kind of injury it was, a household or an industrial injury (they are paid differently)? If it was an enterprise staircase, there is no doubt about the origin of the injury. But if it was a staircase in the house where the worker lives, it all depends on how he went home. If he has not yet reached the door of the apartment from his work, the injury is considered industrial. But if he went to the store or the cinema on the way, it becomes a household injury. Hence, the law defines the limits of the enterprise *conditionally*.

The conditionality of the system's borders again brings us back to the problem of integrity, which now becomes the integrity of the whole world. The system boundary is defined taking into account the goals of the subject who will use the system models.

Structuredness. The *fourth static property* is that parts of the system are not independent and isolated from each other; they are interconnected, and they interact with each other. In this case, the properties of the system as a whole significantly depend on how its parts interact. Therefore, it is often important information on the relationship of the parts. The list of essential links between the elements of a system is called the *model of the system structure*. The endowment of any system with a certain structure is the fourth static property of systems — *their being structured*.

The concept of structuring further deepens our understanding of the integrity of the system: links, as it were, hold parts together as a whole. Integrity, marked earlier as an external property, receives a reinforcing explanation from within the system, that is, through the structure (Figure 2.4).

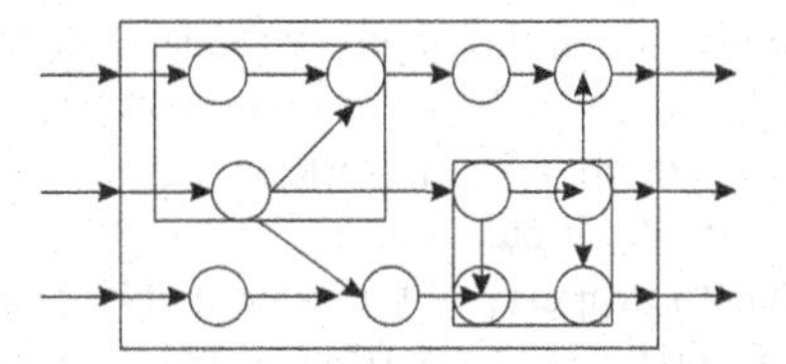

FIGURE 2.4 Ties between parts form a model of system's structure.

If we need to use a model of the structure of the system, we need to take care of the quality of the model, which turns out to be a difficult task.

The difficulties of constructing a model of the structure. We emphasize that for any system many different models of structure can be offered. It is clear that achieving a certain goal will require one, specific, most appropriate model of them. The difficulty of choosing from available models or building a model specifically for our case stems from the fact that, by definition, the structure of the model is a list of *essential* relationships. The word "essential" is evaluative, so its meaning depends on objective circumstances and subjective assessments of these circumstances. Let us discuss the main difficulties that lie in wait for us in determining the structure of the model.

The first difficulty is that the structure model is determined after the composition model is selected, and depends on the composition of the system. But even with a fixed composition, the structure model is variable due to the possibility of differently determining the significance of the connections. For example, a modern manager is advised to take into account that along with the formal structure of his organization (which is defined by statutory documents), there is inevitably an informal structure (due to personal relationships between employees), which also affects the functioning of the organization. To improve management efficiency, one should know and use both the formal and the informal structures. Similarly, in the design and operation of computer systems, one has to simultaneously take into account their hardware and software components on their own, interacting with each other, and sometimes mutually replacing structures.

The second difficulty stems from the fact that each element of the system is a "little black box". Hence, all four types of errors are possible when defining the inputs and outputs of each element included in the structure of the model.

2.2 DYNAMIC PROPERTIES OF THE SYSTEM

Let us consider the second group of system properties called *dynamic properties*.

If we consider the state of the system in another, different from the first, point in time, we will again discover all four static properties. But if you impose these two "photos" on each other, then you will find that they differ in details: during the time between the two moments of observation, some changes occurred in the system and its environment. Such changes can be important when working with the system and, therefore, should be reflected in the description of the system and should be taken into account while working with it. The features of changes with time inside and outside the system are referred to as the *dynamic properties of the systems.* If static properties are what you can see in the photo of the system, then dynamic properties are those that are found when watching a movie about the system. We can speak about any changes in terms of changes in static models of the system. In this regard, four dynamic properties are distinguished.

Functionality is the fifth property of the system. The processes $Y(t)$ occurring at the outputs of the system $\{Y(t) = \{y_1(t), y_2(t),\ldots, y_n(t)\}$, Figure 2.5) are considered as its *functions.*

Functions of the system are its behavior in the external environment, changes made by the system in the environment, the results of its activities, and the products manufactured by the system.

From the multiplicity of outputs follows a plurality of functions, each of which can be used by someone for something. Therefore, the same system can serve different purposes.

A person using the system for his own purposes will naturally *evaluate* its functions and put them in *order* in relation to his needs. Thus, the notions of main, secondary, neutral, undesirable, superfluous functions appear. Note that these terms are evaluative, subjective, and relative. Thus, the main function of a lamp is considered to give light, but when choosing a lamp from dozens of others sold in a store, its decorative qualities, consistency with the interior of the room, its cost, etc. come to the fore.

Therefore, we single out two points of this property of systems: objective plurality of functions and subjective ordering of functions.

Liability to stimulation is the sixth property of the system. At the inputs of the system, certain processes $X(t) = \{x_1(t), x_2(t), \ldots, x_m(t)\}$ also occur, affecting the system, turning (after a series of transformations in the system) into $Y(t)$ (see Figure 2.6). Let us call impacts $X(t)$ *stimuli*, and the susceptibility of any system to external influences and change in its behavior under these influences will be called *liability to stimulation.*

By the way, the word "stimulus" in ancient Greek means a stick for urging on a donkey.

The variability of the system over time is the seventh property of the system. In any system, there are changes that need to be taken into account to foresee and incorporate into the project of a future system; to promote or counteract them; and to

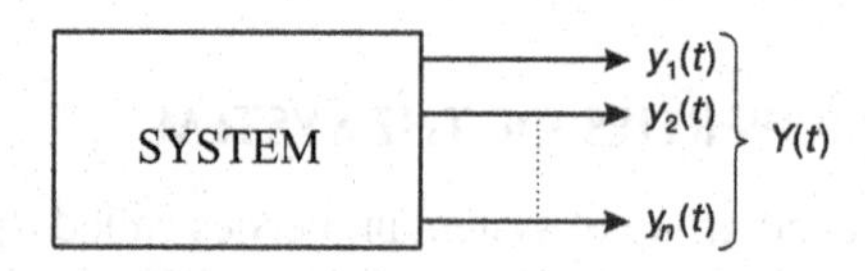

FIGURE 2.5 Processes at outputs are functions of a system.

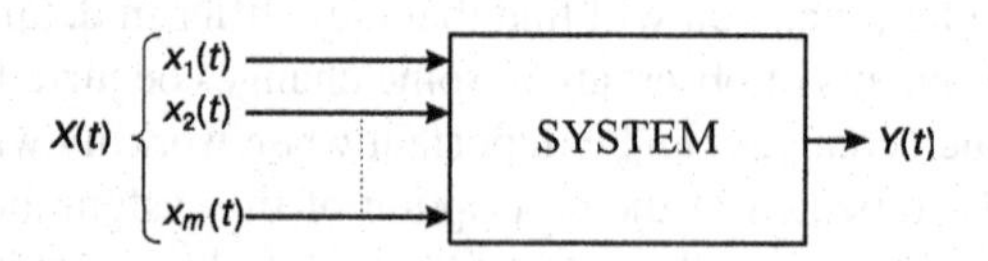

FIGURE 2.6 Processes at inputs of a system are called stimuli.

accelerate or slow them down when working with an existing system. Anything can change in the system, but in terms of our models, we can give a clear classification of changes: the values of internal variables (parameters) $Z(t)$, the composition and structure of the system, and their combinations can change (see Figure 2.7).

The nature of these changes may also be different. Therefore, further classifications of changes may be considered.

The most obvious classification is by the rate of change: fast or slow (compared to something taken as standard); it is possible to introduce a larger number of gradations of speeds (superfast, very fast, etc.).

Of interest is the classification of trends in changes in the system regarding its composition and structure. We begin this classification with the introduction of special concepts, considering the changes in a short time interval, so that the changes can be considered to go in "one way", that is, monotonous.

We can talk about such changes that do not affect the structure of the system: some elements are replaced by other equivalent elements, and parameters (internal variables $Z(t)$) can change without changing the structure (clock, city transport, school, theater, etc., they "work"). This type of system dynamics is called its *functioning.*

Further, changes can be primarily *quantitative*: the number of parts (system's composition) grows, and although its structure changes automatically, for now it does not affect the properties of the system (extension of a landfill or a graveyard are examples). Such changes are called system *growth.*

Then allocate *qualitative* changes in the system, in which there is a change in its essential properties. If such changes go in a positive direction, they are called *development.* With the same resources, the developed system achieves better results, and new positive qualities (functions) may appear. This is associated with an increase in the level of organization of the system.

As applied to organizational systems, R. Ackoff defines development as "an increase in desires and the ability to satisfy one's own and others' needs and justified desires". (Desires are called "justified" if their satisfaction for the sake of some does not negatively affect the development of others. Needs are what is necessary for survival. Different combinations are possible: for example, you may not want the needed, and you may wish the unnecessary).

How much a subject (individual or organization) has is a matter of accumulated wealth. An indicator of wealth is the *standard of living.* But the question of what we can do with what we have is a question of competence, that is, what we have learned.

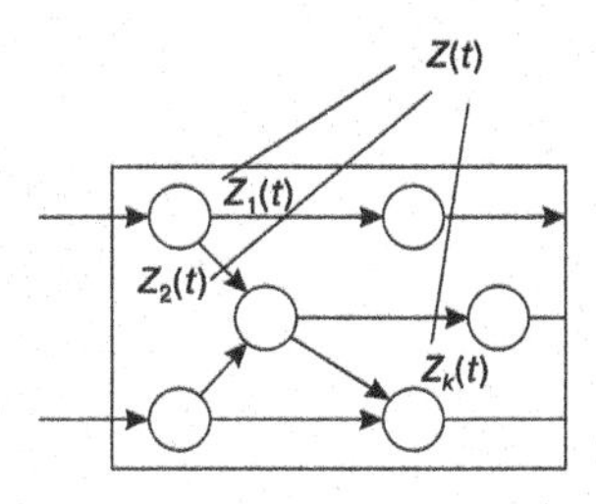

FIGURE 2.7 Changes in the system may consist of changes in composition either in structure or in both.

And this is expressed by the *quality of life* achieved. The more developed is the subject, the less resources he needs to achieve a satisfactory quality of life, or the greater the quality of life he can achieve with what he has.

Therefore, growth occurs mainly due to the consumption of material resources, that is, development caused by the assimilation and use of information. Growth is an increase in the size and number of parts. Development is an increase in competence. Bulk is the result of growth; competence is the result of development. The purpose of the growth of the organization is to improve living standards. The purpose of development of the organization is to improve the quality of life.

Growth may inhibit development, but development cannot restrain growth. Growth and development can go simultaneously (as in a child), but are not necessarily linked. Growth is always limited (due to external physical conditions, in particular, limited material resources), and development from the outside is not limited because information about the external environment is inexhaustible: no matter how much we know, there is always something else that is unknown. Lack of material resources can limit growth but not development.

However, there is an internal constraint on development. The fact is that development is the result of assimilation and use of new information, that is, the result of training is learning outcome. But learning is perhaps the only action that cannot be done for and instead of the learner. If the person does not want to learn, it will not, and cannot, develop. From the outside, it is impossible to develop the system; you can only help in the development, but provided the propensity of the system to learn. ***Development is possible only as self-development***.

It is clear that in addition to the processes of growth and development, reverse processes may also occur in the system. Reverse growth changes are *recession*, contraction, and reduction. Reverse development is called *degradation*, that is, loss or weakening of useful properties.

We have considered the possible monotonic changes of the systems (see Figure 2.8).

Obviously, a monotonic change can't last forever. In the history of any system can be seen periods of decline and recovery, stability and instability, the sequence of which forms the individual *lifecycle of the system*.

The concept of lifecycle deserves special discussion because both in the design of future systems and in the study and management of existing systems, information about the individual history of the system plays a very significant and often decisive role in achieving this goal.

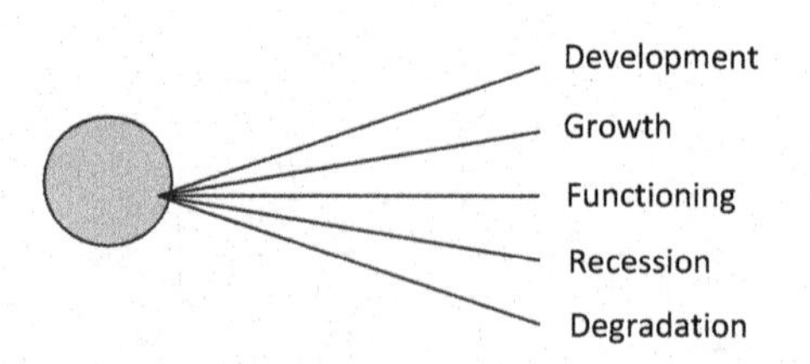

FIGURE 2.8 Types of monotone changes in systems.

When describing the lifecycle, special attention should be paid to the *continuity* of its trajectory. Put differently, it is necessary to determine the lifecycle in the past and the future. The past history is restored according to the information available to us. Unfortunately, often this information is incomplete, inaccurate, and about some periods completely lost. Therefore, the description of past events often inevitably has irreparable gaps (an example is the biography of Jesus Christ contained in the Gospels). But when determining the future lifecycle of a designed system, continuity should be a special concern: the planned history of this system will end at the first gap in the description of its lifecycle. There are many examples of this. The carelessness of the stage of disposal of obsolete fluorescent lamps has led to the fact that mercury enters the soil and water from broken lamps in landfills, poisoning all living beings. The ill-conceived stage of connecting the cables of the space rocket ended with the fact that the ship sent to Mars did not open the antennas because of incorrectly connected signal cable, and all the billions of dollars spent on the project went to waste. How many crop death cases were due to the omission of a single stage of its lifecycle. There should be no gaps in the description of any technology.

Further, while describing the processes occurring in the system, you can use their other classifications, for example, classification by *predictability*: deterministic and random processes. Or classification according to the *type of time dependence*: monotonous, periodic, harmonic, impulse, random, chaotic, etc.

Existence in a changing environment is the eighth property of the system. Changes not only in this system, but in all the rest. For this system, it looks like a continuous change in the environment. The inevitability of existence in a constantly changing environment has many consequences for the system itself, starting with the need to adapt it to external changes in order not to die, to various other reactions of the system.

When considering a particular system for a specific purpose, attention is focused on some specific features of its response. For example, consider how the rate of change within a system should relate to the rate of change in the environment — should it be slower, match, or go faster? This is determined by the nature of the system or its purpose. For example, for systems designed to transfer information in time (books, monuments, works of art, video and audio recordings, triangulation marks, etc.), the slower they change with changes in the environment, the better they perform. Another example of this is the preservation of its state by automata and living organisms (homeostat, stabilization, stationarity). A new reaction of living organisms goes almost simultaneously with changes in the environment, for example, the adaptation of the pupil with changes in lighting. There are systems whose functions can only be performed if changes made in the system are faster than the changes in the environment. Management is a typical example: the search and comparison of various control actions must occur at an accelerated rate so that the selected effect proceeds in real time.

We note another important feature of the existence of the system in a changing environment. The changes themselves are constantly changing; this is reflected in the acceleration of changes in the environment. For example, the speed of movement in space, the transmission and processing of information, and production and consumption of products during our generation has increased more than that in all the prehistory. This requires rapid and significant changes in what we do and how we do it. People, organizations, firms, and governments that are adapting to the changes poorly leave the stage and drop out of the game. The only chance to survive in a turbulent environment is to provide a dynamic balance, like a ship or plane caught in a storm. And the stronger the external changes, the more actively the internal changes should be carried out (compare the driver's activity on good and bad roads, in good and bad weather). Although forecasting and training remain important means, developing immunity to changes beyond our control and strengthening control over others are considered to be more effective.

2.3 SYNTHETIC PROPERTIES OF THE SYSTEM

Now let's move on to the third group of system properties — *synthetic*. This term refers to the generalizing, collective, and integral properties that take into account what was said before, but focus on the interaction of the system with the environment, as well as on the integrity of nature in the most general sense.

Emergence is the ninth property of the system. Perhaps, this property more than all others speaks about the nature of systems. Let's start with examples.

Mechanical example. With two interacting cobblestones, you can produce effects that are impossible when used separately: knocking, carving sparks, pounding nuts, etc.

Chemical example. When hydrogen is combined with oxygen, each having a number of special properties, the H_2O formula creates a new wonderful substance — water. The properties of water, many of which are not fully understood still (the role of water in living and inanimate nature, melt water, magnetized water, with their differences from ordinary water, water memory, etc.), are not derived from the properties of hydrogen and oxygen.

Biological example. Male and female individuals of a bipolar population each have their own individual characteristics. But only at their connection there is an opportunity give birth to new generation, formation of society, etc.

Logical and mathematical example. Let us have two black boxes with one input and one output. Each of them can work only with integers and perform only one simple operation: add one to the number at the input (see Figure 2.9). Let's connect them now in the system according to the ring scheme (see Figure 2.10). We have a system S_1 without inputs and with two outputs.

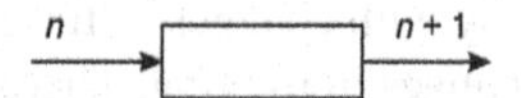

FIGURE 2.9 Separate elements of many similar ones in a system.

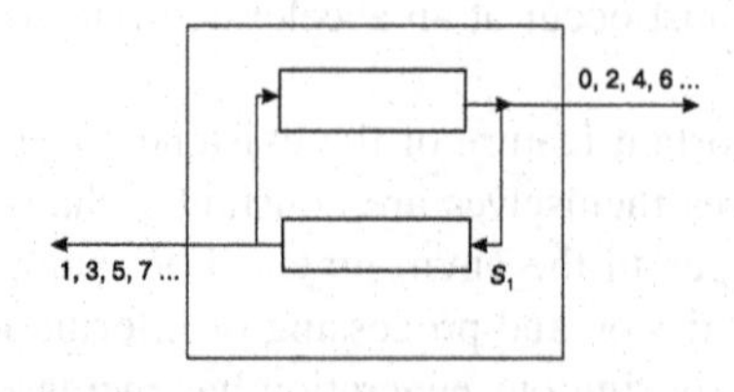

FIGURE 2.10 System S_1 of two elements connected in a closed loop.

At each cycle of operation, the scheme will produce a larger number, and it is remarkable that only even numbers will appear on one output, and odd numbers only on the other output. Isn't that a beautiful example of emergence?

Now we draw conclusions. Combining parts into a system gives the system new qualitative properties that are not reduced to the properties of the parts, are not derived from the properties of the parts, are inherent only in the system itself, and exist only as long as the system is one whole. The system is more than just a collection of parts. The qualities of the system, inherent only to it, are called *emergent* (from the English "arise").

Where do emergent properties come from if none of the parts have them? What is responsible for their appearance in the system? The answer is found in the logical mathematical example. Let's connect the same two black boxes in a different way in parallel (see Figure 2.11).

The resulting S_2 system has one input and one output. If you enter the number n, the output will be $n+1$. It turns out that S_2 is arithmetically identical to each element, and its arithmetic property is not emergent (unlike S_1)! But we already know that the system necessarily has emergent properties. It turns out that S_2 has the ability to perform operation $n+1$ even if one of the elements fails, that is, increased reliability. In reliability theory, this method is known as *redundancy* — improving reliability by introducing redundancy into the system.

It is easy to see that S_1 and S_2, consisting of the same number of identical elements, differ only in the scheme of their connections, that is, in the *structure*. The structure of the system determines its emergent properties.

Let's sum up.

1. The system has emergent properties that cannot be explained or expressed through the properties of its individual parts. Therefore, in particular, not all biological laws can be reduced to physical and chemical; social to biological and economic; computer properties cannot be explained only through electrical and mechanical laws.
2. The source and the carrier of emergent properties is the structure of the system: in different structures, systems formed from the same elements have different properties.
3. The system has also nonemergent properties that are the same as the properties of its parts. For example, for technical systems, it is volume, mass; and arithmetic for S_2, etc. The system on the whole may have nonemergent properties (e.g., car coloring). An important and interesting case where parts

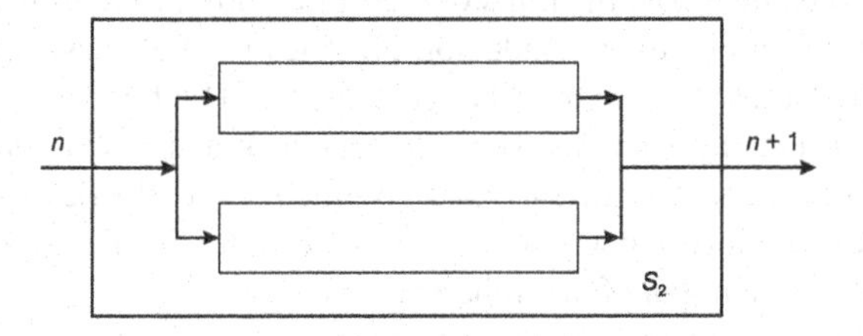

FIGURE 2.11 System S_2 of two elements connected in parallel.

of the system have the properties of the system as a whole is the so-called *fractal* construction of the system. In this case, the principles of structuring parts are the same as that of the system as a whole. Fractals are observed in nature (hierarchical organization in living organisms, the identity of the organization at different levels in naturally growing systems, i.e., biological, geological, demographic, etc.), and mathematicians have developed an abstract theory of fractals.

4. Emergence demonstrates another facet of integrity. The system acts as a whole because it is the carrier of the emergent property: it will not be whole, and the property will disappear; if this property is manifested, then the system is intact. For example, none of the parts of the plane can fly, but the plane flies.
5. Emergence is another more developed form of expression of the law of dialectics on the transition of quantity to quality. It turns out that for the transition to a new quality, the "accumulation" of quantities is not necessary ("the last drop overflowed the bowl", "the last straw broke the back of a camel"). For the emergence of the new quality, it is enough to combine into a whole at least two elements.
6. Note that the dynamic aspect of emergence is denoted by a separate term, *synergy*, and extensive literature is devoted to the study of synergetics.
7. It is interesting to note that while in artificial systems the emergent property arises from the intentional joining of selected parts, in natural systems, emergence determines which parts are to be joined and how they are to interact. So, a living organism defines the meaning of the skeleton, heart, liver, and lungs; a family gives meaning to the roles of husband, wife, and their children. (The emergence of the first is survival in the natural environment, and of the second survival in the social.)
8. The action of the system depends more on how its parts interact than on how they act on their own. Therefore, improving the actions of individual departments of the organization does not necessarily lead to an improvement of the entire organization, and often even vice versa (in mathematics, this is manifested in the difference between local and global optimization; in business, in the form of the expediency of producing unprofitable goods for the sake of increasing the sale of profitable ones).

The main recommendation to managers at any level is that they are not so much engaged in improving the work of individual parts of their unit as in improving the interaction between them and the links of their unit with the environment. An example is the work of an orchestra conductor. He does not tell musicians how to play instruments: they know how to do it better than him. His business is not to control their actions but their interactions. Here, factors such as tracking orchestra players for the actions of others, the presence of a common score, the desire of each to join the harmony, and teamwork begin to play an important role. The work of a leader is even more difficult than that of an orchestra conductor.

Inseparability into parts is the tenth property of the system. Although this property is a simple consequence of emergence, its practical importance is so great, and its underestimation occurs so often that it is advisable to emphasize it separately. If we need the system itself, and not something else, then it cannot be divided into parts.

When a part of the system is removed from the system, two important events occur.

First, it changes the composition of the system, and hence its structure. It will be *another* system, with different properties. Because of the properties of the old system, a lot, some property associated with this part, will disappear (it can be emergent, and not those, for example, compare the loss of phalanx of finger for a pianist, geologist, guitarist, and carpenter). Some property changes, but some will remain the same. Some properties of the system are not significantly associated with the withdrawn part.

We emphasize once again that whether the withdrawal of a part from the system will be significant is the question of assessing the consequences. Therefore, for example, consent of a surgical operation is requested from the patient, and not everyone agrees to it.

The second important consequence of removing a part from the system is that the part inside and outside the system is not the same. Its properties are changed as the properties of the object are manifested in interactions with the surrounding objects, and when removed from the system, the environment of the element becomes completely different. The amputated hand will not grab anything; the pulled out eye will not see anything.

It would, however, be wrong to absolutize the indivisibility of systems. For example, this would mean a ban on surgical operations, or on the organizational transformation of enterprises. It is only necessary to clearly realize that after the separation we are dealing with other systems. This is especially important in the analytical study of the system when its parts are considered separately. Special care is required to maintain the links of the considered part with the rest of the system.

Inherence is the eleventh property of the system. The more the system is inherent, the better it is matched and adapted to the environment compatible with it (inherent means being an integral part of something). The degree of inherence varies and can change (learning, forgetting, evolution, reform, development, degradation, etc.).

The fact that all systems are open does not mean that they are all equally well aligned with the environment. Consider the function "swim in the water" and compare the quality of this function in such systems as fish, dolphin, and scuba diver. They are ordered in the obvious way: fish do not need to leave the aquatic environment at all; dolphin must breathe air; and scuba diver's options are limited capacity of the air cylinder, not to mention the physical and physiological limitations.

The expediency of emphasizing inherence as one of the fundamental properties of systems is caused by the fact that the degree and quality of the implementation of the selected function by the system depend on it. In natural systems, inherence is increased by natural selection. In artificial systems, it should be a special concern of the designer. Illustrative examples include preparation for transplantation of the donor organ and the patient's body, exchange of cultural values, and introduction of technical innovations.

In some cases, inherence is provided by intermediary systems. Let's consider some examples. The hieroglyphic writing of the ancient Egyptians could be deciphered only with the help of the Rosetta stone, on one side of which was the inscription in hieroglyphs (noncoherent to modern culture), and on the other the same inscription in ancient Greek, known to modern specialists. Another example is the adapters to connect European electric appliances to American plug sockets. Another example is the work of an interpreter between two individuals speaking different languages. Medical example: in Tomsk, Professor G. Dombayev developed a method for treating diabetes by transplanting the cells of the calf gland to a patient. But the human body quickly detects alien (noncoherent) implant and rejects it. However, for some reason, the body does not reject some metals (war invalids sometimes live their lives with metal fragments in the body). The solution was found in implanting a porous metal capsule with the cells of someone else's healing gland into the patient's body.

The problem of inherence is important in all cases of working with systems. Striking examples are management and leadership (compatibility of the head with the supervised ones), marketing and innovation (the inherency of the proposed product to the target consumers), pedagogical skills (coordination of the teacher with the audience), standardization service (care about the compatibility of products produced at different enterprises), training of illegal spies (ensuring their indistinguishability from the citizens of the country under investigation), etc.

In conclusion, we emphasize that inherence is not an absolute property of the system and is tied to a specific function. In particular, if we take our example of a fish, a dolphin, and a scuba diver in water and consider the same situation with respect to the function "to carry out electric welding under water", then these three systems will be ordered by the inherency in a completely different order.

Purposefulness is the twelfth property of the system. In man-made systems, the subordination of all (both composition and structure) to the goal is so obvious that it should be recognized as a fundamental property of any artificial system. Let's call this property *purposefulness.* The purpose for which the system is created determines which emergent property will ensure the realization of the goal, and this, in turn, dictates the choice of the composition and structure of the system. One definition of the system is that *the system is a means to an end.* The implication is that if the proposed goal cannot be achieved at the expense of the existing opportunities, then the subject composes from the surrounding objects a new system, specially created to help achieve this goal. It is worth noting that rarely the goal uniquely determines the composition and structure of the created systems: it is important to implement the desired function, and this can often be achieved in different ways. At the same time, attention is drawn to the similarity of the structure of different representatives within the same type of systems (living organisms, vehicles, planetary systems, mineral deposits, etc.).

The problem of purposefulness in the nature. Turning to the natural objects, we find that they have all the previous 11 properties of systems, and often the manifestation of these properties is much more sophisticated than that of artificial systems. There was even a special science of bionics developed, "spying" the secrets of harmony and perfection of living organisms in order to transfer the principles found into the technique. In the inanimate nature, there are obvious manifestations of systemic character: physical, chemical, geological, astronomical objects on all grounds should be attributed to the systems. Except but one: purposefulness.

The first way out is to draw an analogy between artificial systems and natural objects. This analogy identifies artificial and natural systems and makes us look for a goal-setting subject outside the universe itself. At the same time, we have to admit that the intellect of the Creator is incomparably superior to the human mind. This is the basis of the emergence of religions. The question naturally arises: is God himself a system? Different religions have different views on this issue. Some declare it to be meaningless as the human mind cannot know the complexity of the Creator that exceeds its capabilities; it is proposed to believe that He is His own cause and effect. There are, however, religions that do not consider this question heretical; they propose the hypothesis of the hierarchy of deities: there are gods for people, then there are gods for the gods of people, and so on to infinity.

However, it is possible to propose another hypothesis about the similarity, but not the identity of man-made and natural systems, which allows to solve the arisen difficulty without requiring a mental exit from the limits of the universe. For this, it is necessary to clarify and specify the concept *of purpose*.

What is the purpose? Let's see how the concept of the purpose develops, deepens, and clarifies on the example of the concept of an artificial system close to us.

The history of any artificial system begins at some point 0 (Figure 2.12) when the existing state of the outputs $Y(t = 0) = Y_0$ turns out to be unsatisfactory, that is, there is a *problem situation*. The subject is unhappy with this condition and would like to change it. When asked what he would like (what is his purpose), he answers that he would be satisfied with the state of Y^*. This is the starting point for defining the purpose. Further, it is discovered that Y^* does not exist now and also cannot be achieved for a number of reasons in the near future. The second step in defining the goal is to recognize its desired *future* state. Immediately it turns out that the future is unlimited. The third step in clarifying the concept of the goal is to estimate the time T^* when the desired state Y^* can be achieved under given conditions. Now the ultimate purpose becomes two-dimensional, it is a point (T^*, Y^*) on our chart.

The task now is to move from point $(0, Y_0)$ to point (T^*, Y^*). But it turns out that this path can be followed by different trajectories, each of which begins at $(0, Y_0)$ and ends at (T^*, Y^*), and only one of them can be realized. There is a problem of comparison and choice of the best trajectory. Let the choice fall on the trajectory $Y^*(t)$.

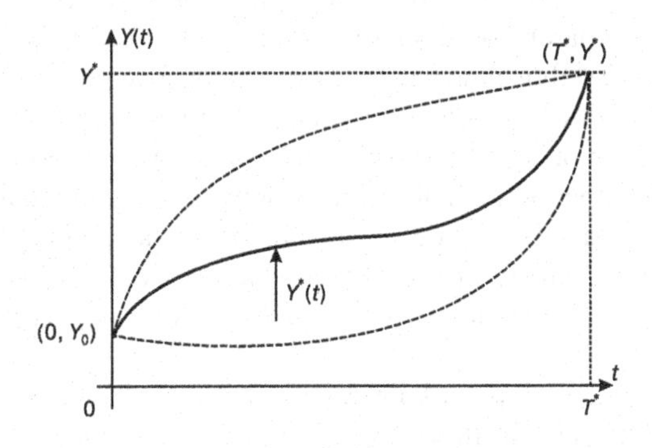

FIGURE 2.12 Many trajectories between problem and purpose, and the best one.

This means that it is not only desirable for us to arrive at point (T^*, Y^*) but to arrive through a sequence of states on the curve $Y^*(t)$. Thus, it is necessary to include into the notion of purpose all the desired future states, as well as the final and intermediate ones. This is the fourth and the final step in defining the *purpose*: the purpose now means not only the final state (the end) (T^*, Y^*) but the entire trajectory $Y^*(t)$ (intermediate objectives, goals, aims, plans, auxiliary operations, etc.).

So, ***the purpose of the artificial system is all its desired future history, $Y^*(t)$.***

The purposefulness of natural objects. Now let's look at our chart from a different point of view. Looking at (T^*, Y^*) from position $t = 0$, we consider it a desirable future state. After time T^*, this state becomes real and can be achieved by the present. Therefore, it is possible to define the *final goal as the future real state*. This is a decisive step toward the interpretation of purposefulness in nature: after all, any, including natural, object is sure to come in the future to some state. This, by definition, is the purpose. Importantly, we don't need to hypothesize about someone setting a purpose in advance.

Now we have the opportunity to say that the property of purposefulness is present in natural systems equally as the artificial ones. This makes it possible to approach any system from a single point of view and with a single methodology.

To dispel the confusion that arises, we honestly and clearly recognize that "the goal as the image of the desired future" and "the goal as the real future" are not the same. Let us introduce different terms for them: the first will be called the *subjective* goal, and the second the *objective* goal.

This, first, clarifies the difference between artificial and natural systems: artificial systems are created to achieve subjective goals; natural systems, obeying the laws of nature, realize objective goals.

Second, it clarifies the reason that not every subjective goal is achievable. The fact is that not only the lack or misuse of available resources can be the cause of failure. The main condition for achieving a subjective goal is its belonging to the number of objective goals: only goals that can become a reality are feasible.

One of the reasons for the emergence of unattainable subjective goals is that subjective goals are the product of imagination, and objective ones are the result of the manifestation of the laws of nature. Restrictions on mental constructions are much weaker than those on possible real events. You can imagine a beautiful creature — a half-girl-half-fish — a mermaid, or no less handsome creature — half-man-half-horse — centaur, but nature does not allow their appearance.

It is important to establish the feasibility of a subjective goal before attempting to realize it. Unwillingness to waste efforts and resources would allow not to engage in the implementation of an unattainable goal. So far, we have only one criterion of unattainability — a contradiction to the laws of nature (e.g., the purpose of creating a perpetual motion machine). However, sometimes we cannot cite the laws of nature that impede the achievement of a goal (e.g., the goals of creating artificial intelligence; and although success in this is far from expectations, efforts do not seem to be in vain).

There is, however, one type of knowingly unattainable goals that are not considered unworthy of striving for them. Such goals are called *ideals*. The peculiar feature of the ideal is that although it is obviously unattainable, still it is attractive, and most importantly, allows it to be approached. For example, harmonically developed personality; the desire to go on increasing sports achievements; the knowledge of an increasing number of languages; in general, the pursuit of excellence in any respect.

2.4 CONCLUSION (SYSTEMS PICTURE OF THE WORLD)

In this chapter, an attempt is made to represent the world as a world of systems interacting with each other, entering as parts into larger systems, containing smaller systems, each of which continuously changes and stimulates other systems to change.

From an infinite number of properties of systems allocated, 12 are inherent in all systems. They are allocated on the basis of their need and sufficiency for justification, construction, and accessible exposition of technology of the applied systems analysis.

However, it is very important to remember that each system is different from all others. This is manifested primarily in the fact that of each of the 12 system-wide properties, these properties are embodied in an individual form specific to each system. In addition to these system-wide patterns, each system has other specific inherent properties.

Applied systems analysis is aimed at solving a specific problem. This is reflected in the fact that using the system-wide methodology, it is technologically aimed at the detection and use of individual, often unique features of a problem situation.

To facilitate such work, you can use some classifications of systems, fixing the fact that different systems should use different models, different techniques, and different theories. For example, R. Ackoff and D. Gharayedaghi [4] proposed to distinguish systems by the ratio of objective and subjective goals of the parts of the whole: technical systems, man-made machines, social, and environmental. Another useful classification, according to the degree of knowledge of the systems and the formalization of models, was proposed by W. Checkland [5]: "hard" and "soft" systems and, accordingly, "hard" and "soft" methodologies, as discussed in Chapter 1. M.C. Jackson [6] describes 10 variants of problem-solving technologies according to differences between problem situations, such as level of disagreement between its participants, and its position between hard and soft modeling.

Thus, it can be said that the systems vision of the world is to begin to consider a particular system, focusing on its individual characteristics and understanding its universal systems nature. Classics of systems analysis formulated this principle aphoristically:

"Think globally, act locally".

QUESTIONS AND TASKS

1. What are the static properties of systems? List four static properties.
2. How does the fact of universal interconnectedness in nature follow from the openness of systems?
3. What is the black box model called? What are the four kinds of mistakes you can make when building a black box model?

4. What is a system composition model? What are the (three) difficulties of building it?
5. Under what assumptions can we talk about the presence of parts of the system?
6. How is the system boundary determined?
7. What is a system structure model? What are the difficulties of its construction?
8. What are the dynamic properties of systems? List them (all four).
9. Explain the difference between the growth and development of the system.
10. What do we call synthetic properties of systems? List four such properties.
11. Which of the static properties of the system ensures the existence of emergent properties of the system?
12. What is a subjective goal?
13. What is meant by the objective purpose of the system?
14. Why are none of the subjective goals achievable?
15. Define the following concepts:
    - system integrity;
    - openness of the system;
    - black box;
    - error of the first (second, third, fourth) kind;
    - system composition model;
    - subsystem;
    - system element;
    - system structure model;
    - system function;
    - stimulation of systems;
    - functioning;
    - growth (decline);
    - development (degradation);
    - lifecycle;
    - emergence;
    - inherence;
    - the subjective purpose;
    - the objective purpose.

3 Models and Modeling

3.1 MODELING IS AN INTEGRAL PART OF ANY ACTIVITY

The subject, existing in the real world, interacts with it and carries out this or that activity. All possible activities can be divided into two types: knowledge of the world and its transformation (Figure 3.1). It is important to understand that any activity of the subject becomes possible only through models – systems, the specificity of which is aimed at ensuring interaction between the subject and the reality. Models play the role of an intermediary between them.

Modeling is not an action that you can do or cannot do: modeling is an inevitable, indispensable, and mandatory part of any human activity (not only humans, but this is a separate topic). We show this for both types of activities.

Let's start with the transformation activity. Whatever a person does toward changing reality, even before the work itself begins, he must determine the ultimate purpose, the *image* of the desired future, that is, a *model* of what is not yet present, but what would be accomplished, what would appear at the end of the work. This is the first argument about the need for modeling. But that's not all: to achieve the final result, you must perform a certain sequence of intermediate actions, but to perform them correctly, you need to describe this sequence even before the work begins, that is, create a *model* (plan, algorithm) of the progress. Thus, transformative (labor, managerial) activity is impossible without modeling.

Now let's turn to cognitive activity. The final result of knowledge — information about the environment — must be recorded, described, and presented in the form of a specific model. Knowledge generally exists only in the form of models, that is, ***the***

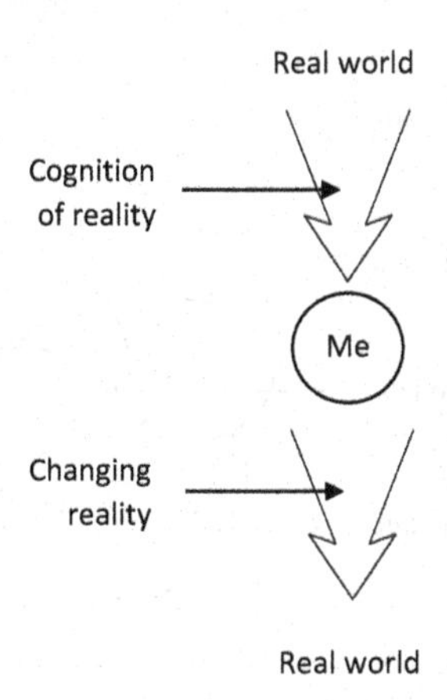

FIGURE 3.1 Two types of interaction between a subject and the reality.

model is a form of knowledge existence. The ultimate goal of knowledge is to build models of the part of the world we are interested in.

Therefore, knowledge is impossible without modeling. It is also important that the process of obtaining information from the outside takes place with the help of special models. It appears that it is necessary to open your eyes and the information will flow through them. All languages differ in the concepts of "to look" and "to see". You can look and not see, and you can see what is not really there. A doctor examining a patient pays attention to such external signs that do not say anything to the layman. Using hypnosis or using the effects of optical illusion can make a person see a non-existent.

All this is because the information brought by light signals is processed by our inherited models before reaching a state that is perceived as a visual image. This applies not only to vision but also to all other senses, which are the channels of communication of the subject with the environment.

It should be recalled that the subject is also an object. The object "directly" interacts with the environment, and the subject interacts only through models. For example, touch the hot plate, you will burn as an object. But if you block your models with hypnosis or drugs, you won't feel anything. Even more evident is the well-known psychological experience, when a person is convinced that the iron is hot, there is a burn blister after touching it, although in fact the iron was cold.

So, any activity of the subject or any interaction with the outside world occurs through models, although as an object it is directly related to the environment (Figure 3.2).

The world of subject models begins to grow on the basis of innate models by extracting information from the experiences of life. Since the innate models of different individuals are different (especially for those born with defects of senses), and personal life experience is purely individual, the world of models built to date in

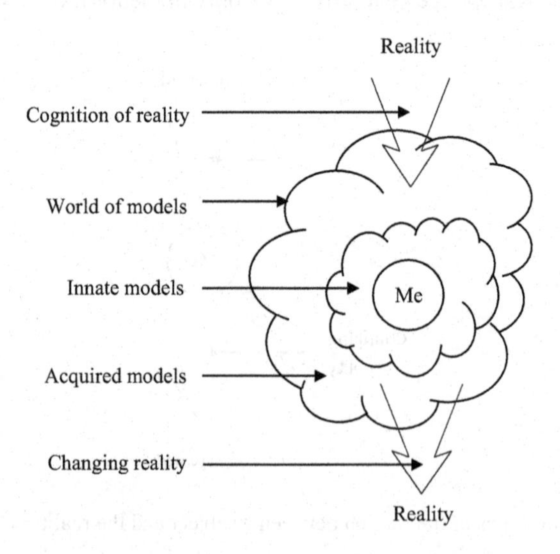

FIGURE 3.2 Interaction with reality occurs with the help of models.

each subject is unique. This implies that everyone sees, perceives, and evaluates the world in their own way. (Of course, this does not mean that there is nothing in common between the subjects; on the contrary, many of their models can be identical or compatible.) This diversity of individual worlds plays a significant role in society and must be taken into account when working with people.

Having established the extreme importance of modeling in the life of the subject, we proceed to consider how models are built, and then to discuss the important properties of various models.

3.2 ANALYSIS AND SYNTHESIS AS MODEL BUILDING METHODS

We need to know and understand some complex systems, that is, to translate them from the difficult and obscure to simple and understandable. This means that we should build a model of this system containing the information we need. Depending on what we need to know, and to explain how the system is arranged or how it interacts with the environment, there are two methods of cognition: (1) analytical and (2) synthetic.

The analytical procedure consists of three consecutive operations:

1. the complex whole is divided into smaller parts, presumably more simple;
2. trying to give a clear explanation of the received fragments;
3. combining the explanation of the parts into the explanation of the whole.

If some part of the system at the second stage is still incomprehensible or unclear, the decomposition operation is repeated and we try again to explain the new, even smaller fragments (Figure 3.3). In the figure, the explained objects are shaded. (In some cases, the analysis of a single branch may be lingered without reaching the explainable fragment. This is a sign of the lack of knowledge that can make a fragment elementary. Positive knowledge in this case is the discovery of what kind of knowledge we lack.)

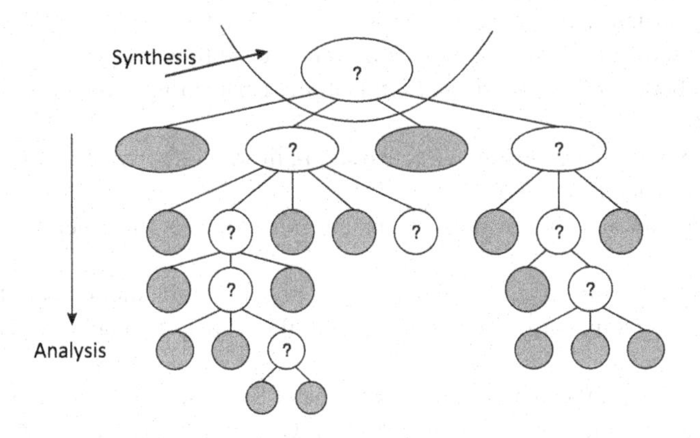

FIGURE 3.3 Analysis begins with modeling parts of the system.

The obtained knowledge is presented in the form of models of our system.

The first product of the analysis is, as can be seen from the scheme, a list of system elements, that is, *a model of the system composition*.

The most serious analysis trap is the danger of breaking the connections of the parts during decomposition, thereby destroying the emergent properties of the system. Therefore, the correct, qualitative analysis needs to realize the distinction between parts, rather than breaking apart the system to independent fragments. Otherwise, it will be impossible to perform the final operation in the analysis: the explanation of the whole is impossible only through the explanation of the parts. To explain the whole means to reveal its emergent properties, and for this it is necessary to establish (or restore) connections between the parts. Thus, the second product of the analysis is *the model of the system structure*. The third product of the analysis are *black box models* for each element of the system.

Thus, from the analysis, we get information about the construction and operation of the system. All received information is presented by all three types of models: composition, structure, and black box.

The analytical method has given remarkable results concerning human cognition of the world. The entire structure of our knowledge is hierarchical: a single world is divided into separate areas chosen by the subject of research, by different sciences: physics, chemistry, history, etc. Each science has its own analytical organization. In each field of knowledge, the matter is brought to the elements from which all the objects of research are formed: elementary particles in physics, molecules in chemistry, phonemes in speech and symbols in writing, cells in biology, notes in music, etc. The success of the analytical method is so significant that there is even the impression that it is the only scientific method (often in common language the words "study" and "analyze" are used as synonyms).

However, there are questions that cannot be answered by the analysis in principle, since the answer does not lie in the internal organization of the system.

Try by any (chemical, physical, artistic) analysis to find out what is the strength and value of the banknote. You can study human anatomy in depth, but it does not explain why nature created two genders. It is possible to investigate in detail the construction of clocks and watches, but it will not give the answer why they are necessary. The study of the structure of a car will not answer why England adopted left-hand traffic.

Answers to questions of this kind lead us to the synthetic approach.

The synthetic method consists in sequentially performing three operations:

1. selection of a larger system (metasystem), in which the system of interest to us is included as a part;
2. consideration of the composition and structure of the metasystem (its analysis);
3. an explanation of the roles that our system plays in the metasystem through its connections with other subsystems of the metasystem (Figure 3.4).

The final product of the synthesis is knowledge of the connections of our system with other parts of the metasystem, that is, *its black box model*. But to build it, we need to create *models of the composition and structure* of the metasystem as by-products.

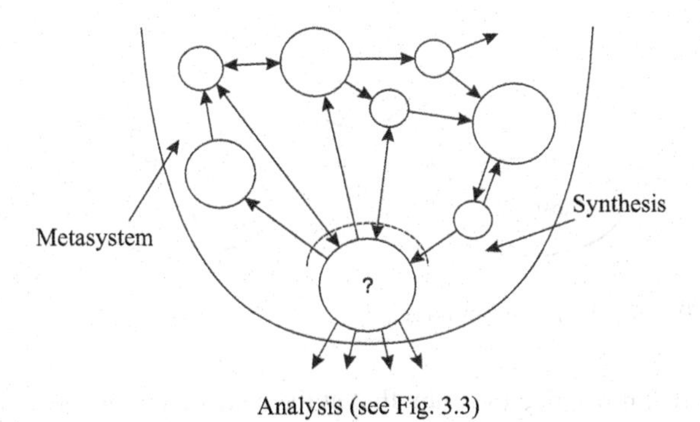

FIGURE 3.4 Synthesis starts with analysis of the metasystem.

Again we see that all the knowledge we have gained is "packed" into three well-known models: the black box, the composition, and the structure. It is clear that the quality of the synthesis directly depends on the quality of the model of the metasystem, which should be especially taken care of.

Analysis and synthesis are not opposites, but complementary. Moreover, there is a synthetic component in the analysis, and a metasystem analysis in the synthesis. What of them or in what sequence to apply them in a particular case is up to the researcher.

3.3 WHAT IS A MODEL?

We have already formulated two definitions of the model. First, *the model is a means of carrying out any activity of the subject.* Second, *the model is a form of the existence of knowledge.* It is possible to somewhat complement each of these definitions by indicating that the model is also a system, with all the system-wide properties described in Chapter 2. A distinctive feature of models from other systems is (in addition to what the two definitions say) in their ability to display the simulated original and replace it in a certain respect, that is, contain and present required information about the original. Let us express this idea in the form of one more general definition: *the model is a systemic representation of the original.*

All three definitions are very general in a philosophical sense. Further, we will need to specify the types of models and their characteristic properties. As we already know, the description of the model can be refined using analysis and synthesis.

3.4 ANALYTICAL APPROACH TO THE CONCEPT OF A MODEL

The analytical approach is aimed at finding out what the system under consideration consists of. We will do the first decomposition, noting that there are two types of materials from which models are created: means of thinking and material means — objects and substances. Accordingly, the models are divided into *abstract* and *real* (Figure 3.5).

We will continue the analysis and carry out the decomposition of the branch "abstract" models. What and how are they built? The answer to this question would mean that we would explain how thought processes occur. But thinking is so

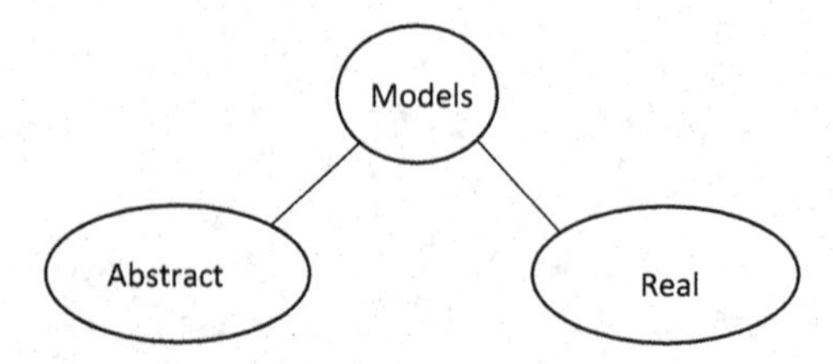

FIGURE 3.5 Two types of abstract models.

complex, so little known phenomenon that at the next step of decomposition we have mostly unexplained fragments (Figure 3.6): most of the manifestations of thinking, noted in our scheme (and not exhaustive of its manifestations), defy explanations that meet the scientific standards of completeness.

More specifically, we can talk about abstract models, which are the endproduct of thinking, prepared for transfer to other subjects. These are models embodied by means of language: they are amenable to registration (texts, audio recordings) and can be studied as objects alienated from direct thinking, but its products contain information about it.

The role of language in society is impossible to overestimate: it is not only a means of communication but also a carrier of culture, a means of organization and governance, and the main component of subject's models of the world. Linguistics is one of the most complex natural sciences (which for some reason is attributed to the humanities).

We will pay attention to those features of the language that will be required to substantiate and use the technology of applied systems analysis.

The main feature for us is that language is a universal modeling tool: you can talk about anything. Of the many properties of language that provide this property to it, let us pay attention to the *vagueness of the meaning of words*.

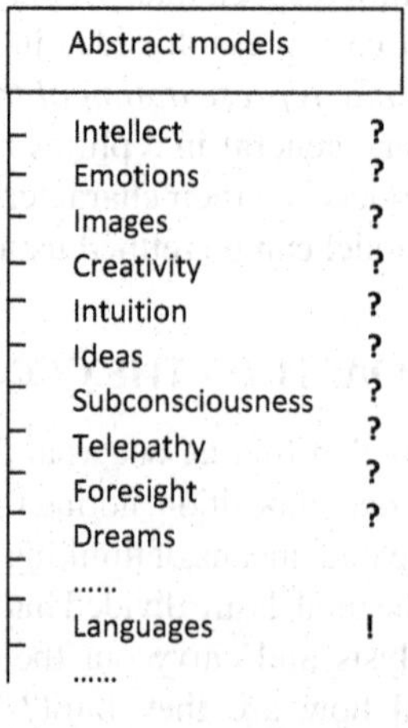

Abstract models

- Intellect ?
- Emotions ?
- Images ?
- Creativity ?
- Intuition ?
- Ideas ?
- Subconsciousness ?
- Telepathy ?
- Foresight ?
- Dreams ?
- ......
- Languages !
- ......

FIGURE 3.6 Among abstract models, only languages are more or less enough studied.

Let us consider an example of a verbal model of a certain situation. "A tall handsome young man entered the room, carrying a heavy burden". You easily can imagine the real situation. But "tall" — what is his height? "Young" — how old is he? Not to mention what "beautiful" is. "Heavy" — what weight? Virtually, no word of a colloquial language has an exact meaning.

An analogy can be made: the "meaning" of a particular situation is a point, the "meaning" of a word is a cloud. Describing a specific situation, we seem to cover the point with clouds, realizing that the truth is somewhere in the middle of this cluster.

In most cases, especially in everyday life, such an approximate, vague description is enough for successful activity. In some types of activity, such vagueness is consciously used as an important positive factor: poetry, humor, politics, diplomacy, fraud, etc.

However, in cases where it is necessary to produce a specific product and to achieve a specific result, this specificity begins to interfere with the vagueness of the everyday language. In such a scenario, those who are engaged in a specific activity get rid of interfering uncertainty by introducing more precise terms into the language. Every group, with its common goal, develops its own, specific language, providing the necessary precision for this activity. The Masai pastoral tribe in Africa has hundreds of terms to describe cows; among the northern peoples, many terms define the condition of the snow; physicists, doctors, lawyers, and criminals speak their special languages; young people talk in slang that is not understandable by adults; London "lower classes" talk on "Cockney".

The general conclusion is that any group activity requires the development of a special, more accurate spoken language; let's call it *professional.*

Professional languages are more accurate than colloquial because of the greater certainty of their terms. It is important to realize that uncertainty can be removed only at the expense of new, *additional information*.

Thus, the accuracy of the meaning of language models is increased due to the acquisition and inclusion in the language of more and more new information about the subject of interest.

Is there a limit to this refinement process? There is, and this is the language of mathematics where terms are as precise and unambiguous as possible. True, it is impossible to completely eliminate uncertainty; otherwise, it would be impossible to talk about the infinity of the world with finite phrases. There are several (and not only auxiliary, but also basic) concepts in mathematics that have a vague meaning: "approximately equal", "significantly more (less)", "infinitely small (large)", "indefinite", etc. Yet, the mathematical language is extreme and is the most accurate on the right end of the spectrum of languages (Figure 3.7).

Now we can appreciate the words of Kant, repeated and interpreted by many philosophers, that the more a field of knowledge can claim to be a science, the more it uses mathematical models.

The fact that some kind of science is insufficiently mathematized (history, biology, medicine, psychology, political science, etc.) means that its object is so complex and so little studied that it is still far from mathematical accuracy. But there is a prospect.

For completeness, we note another feature of languages. The culture of an individual (the world of his models) is formed from the interaction of his innate models

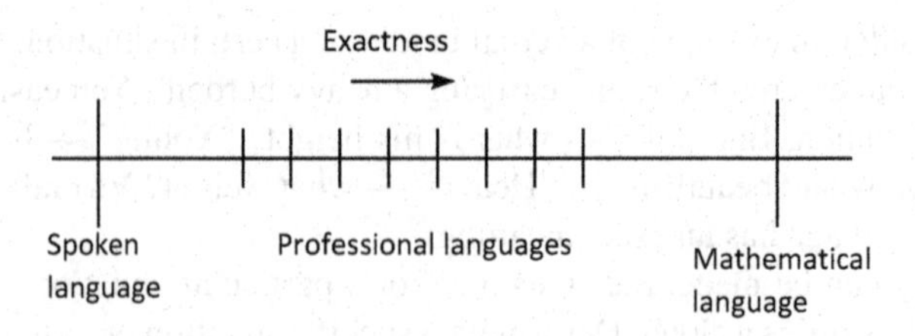

FIGURE 3.7 Spectrum of languages for specific types of activities.

and the culture of his social environment, including (and perhaps first of all) the languages that make up this culture. This is where the inherent property, the compatibility of internal and external cultures, is manifested. At times, they cannot be fully coordinated: genetics have proven that sometimes the inability to learn to speak and write correctly is inherent in the subject's genes; a person who does not have an absolute hearing, despite all his efforts, cannot fully understand the language of music; people with predominant "right-hemispheric" thinking experience a kind of allergy to mathematics; people who matured into emigration experience a feeling of nostalgia; etc. However, there are so many languages that everyone finds an opportunity to shape their culture in a way that in some respects they can ensure their feasible successful interaction with the environment.

The following two points are also important:

1. there is a range of languages of varying degrees of certainty and, therefore, it corresponds to a range of models of varying degrees of accuracy;
2. one of the main features of the applied systems analysis is an attempt to develop the description of the problem situation, presented by the client, in the direction of more and more accurate description of the problem situation from its original "soft" description toward the "hard" one.

Often the mathematical description is not necessary: the problem can be solved without precisely defining the conditions. However, the tendency to include more and more information in the model until it is sufficient to solve the problem is important; the fact of movement toward clarification is important.

We now move on to the next act of decomposition, trying to get to the elementary abstract models. Obviously, *words* are among the elementary language models. What is the original for a word which is its model?

3.5 CLASSIFICATION IS THE SIMPLEST ABSTRACT MODEL OF THE DIVERSITY OF REALITY

Perhaps the main purpose of language is to present finite description of the infinitely diverse world. There are no absolutely identical objects in the world. Even if, for example, electrons are identical in their electrical and mechanical properties (although there is reason to say that this is not quite so: the law of their distribution, derived from the assumption of their identity, is not exactly fulfilled in the experiment), they differ at least by coordinates. Moreover, modern physics prefers to model electrons not as particles but as waves.

How to describe an infinitely diverse world with finite phrases? There is only one way, to do it coarsely, approximately, and in a simplified manner.

The first simplification step is based on the fact that all objects are different, but some differ from each other "weakly", "slightly", or "insignificantly", while others "strongly" or "substantially" (pay attention to the evaluating meaning of these terms). The idea now is to unite all the different but similar objects into one group, leaving others that are considerably different out of it.

The second simplification step is to refuse to take into account the differences within the group, neglect the small differences, and consider the members of the group as the same. This group is called *the class.*

The objects that remain outside the class are also diverse, and although they are "very" different from those that are included in the class, according to other features, they turn out to be "similar" or "different". This makes it possible to identify new classes that are similar inside them and different from other classes.

Consequently, an infinitely diverse world is described by a finite set of classes differing from each other.

To express differences between classes, they are assigned different names (nouns, symbols, signs, numbers, etc.). These names are the words of a certain language. Trees (all trees), animals (all animals), people, buildings, insects, rivers, etc. are examples of class names. Not only objects can be classified but also properties (colors, sounds, forces, dimensions, etc.) and processes (to walk, run, pull, eat, drink, etc.). Thus, the words of the language are the names of some classes.

Classification is the simplest abstract model of the diversity of reality.

Recognition and identification of an object in this case involve finding out which class it belongs to (and what name it should bear).

When building models, the subject has room for informal, creative actions.

First, it is necessary to choose a characteristic, a parameter, and a measure of the difference between objects. The multiplicity of characteristics is one of the reasons for the multiplicity of classifications. A special (and nontrivial) question is the classification not by one but by several characteristics.

Second, the number of classes to be distinguished and the definition of the boundaries between them depend on the specification of the evaluative concepts of "weak" and "strong" differences.

It is important to remember that any classification is only *a model* of the diversity of reality, that reality is more complex, and that there is always an object that cannot be unambiguously attributed to a particular class.

3.6 ARTIFICIAL AND NATURAL CLASSIFICATIONS

There are two types of classifications: *artificial* and *natural.*

In artificial classification, the division into classes is made "as is necessary", that is, based on the purpose,—as many classes and with as many boundaries as is dictated by the purpose of modeling. For example, peasant families in the 20s of the 20th century in Siberia they differed in prosperity q according to the "bell-shaped" distribution (Figure 3.8). For some purposes, the division of peasants into three classes was introduced: poor, middle-class, and reach kulaks, which oversimplified the description of their diversity. On the basis of this model, the Bolsheviks set the task of "eliminating kulaks as a class" and realized this goal. It is characteristic that the boundaries between the classes were not clearly defined, which only increased the injustice. No wonder artificial classification is also called *arbitrary*.

A slightly different classification is achieved when the considered set is clearly inhomogeneous (Figure 3.9). Natural groupings (called clusters in statistics) seem to suggest being defined as classes, as shown in Figure 3.9 (hence the name of the classification *natural*).

However, it should be borne in mind that the natural classification is also only a simplified, coarsened model of reality. For example, the seemingly obvious division of objects into "living" and "dead" faces difficulties in determining the legality of the removal of organs from the deceased person for their transplantation to the living: it is not always obvious that the victim cannot be returned to life. Another example is the "obvious" division of people into men and women, as well as those born as hermaphrodites; sometimes (according to statistics 4%) individuals, because of the entanglement in their bio-chemo-psychological processes, are not able to uniquely identify their own sex. The Olympic Committee had to introduce a genetic test in a female power sports as one of the absolute champions of the world was found to be a man (with all the external signs of a women, though crude).

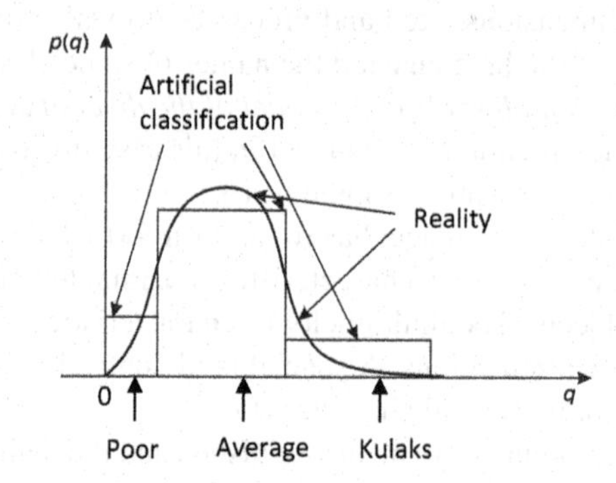

FIGURE 3.8 Artificial, or arbitrary, subjective classification.

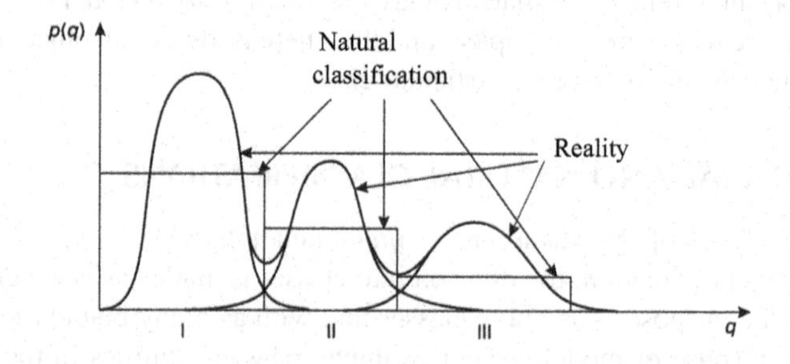

FIGURE 3.9 Natural or objective classification.

As the simplest model, classification underlies other, more complex abstract models. This is achieved both by increasing the number of classes and by introducing more and more relationships between classes.

In some cases, the disadvantages of unambiguous classification become unacceptable. Two types of classification generalization are developed: statistical and vague.

When classifying random objects or quantities, the concept of overlapping distributions is introduced and classification errors are associated with this overlap. For example, in Figure 3.10, the boundary C divides the set X into two classes, X_1 and X_2, associated with density distributions, $p_1(x)$ and $p_2(x)$. (The shaded regions equal the probabilities of errors.)

Another type of classification uncertainty is described by the theory of vague (fuzzy) sets. This theory is based on the assumption that one object belongs to different classes at the same time. In this model, there is no clear boundary between classes. We can only talk about the degree of belonging of the object to a particular class. This degree is expressed by the *membership function*, which takes a value from 0 ("certainly does not belong") to 1 ("certainly belongs"). For example, consider the classification of numbers into "small", "medium", and "large", a number can belong to all classes simultaneously, although to different degrees (i.e., with different values of the membership function $N_A(n)$) (Figure 3.11).

This concludes the consideration of abstract models as limited knowledge of them is enough for us to present the subsequent material.

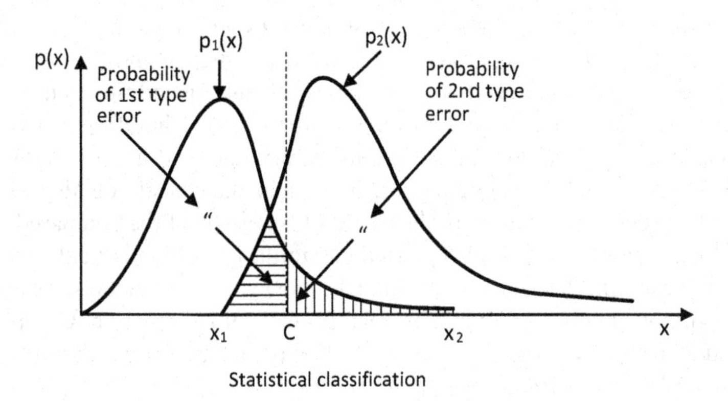

FIGURE 3.10 Statistical classification.

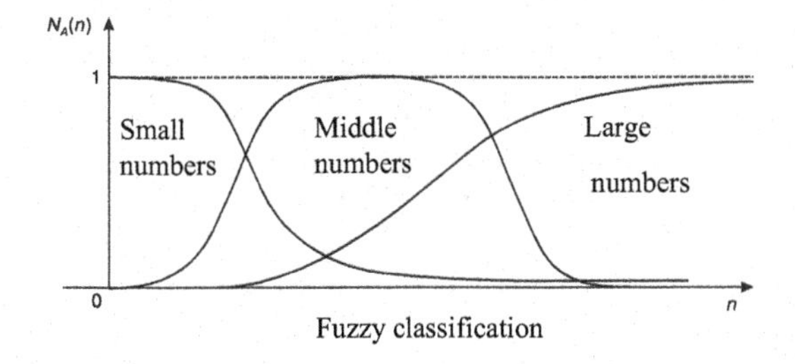

FIGURE 3.11 Fuzzy classification.

3.7 REAL MODELS

The second class of models form real objects used as models of considered system. The analytical technique of classification by the origin of similarity between the model and the original leads to three types of real models.

The first type is called *direct similarity models*. Direct similarity between the model and the original is established by their direct interaction (traces, fingerprint, printing, etc.) or due to a chain of such interactions (photo, building layout, toys, etc.).

The second type is the *models of indirect similarity*, or analogy. Analogy, the similarity of the two phenomena, is explained by the coincidence of the natural laws they obey. Abstract models (theories) of two phenomena can "overlap", which leads to the similarities of these phenomena. Therefore, by watching one of them, you can make a judgment about the other (see Figure 3.12: O — "object", M — "model"). For example, consider the electromechanical analogy, Newton's law $F = m \bullet a$ and Ohm's law $U = R \bullet I$ are structurally identical. This allows us to model mechanical systems by electrical ones, which are more convenient to work with. In many buildings and structures (bridges, towers), piezoelectric sensors are connected to the electrical model of the structure. This allows you to judge its condition and decide on its maintenance. Other examples of analogies include subordination to Kirchhoff's law of currents in power networks, water flow in pipelines, information in communication networks, and transport flow on city streets. It is possible to work out optimal structures and processes for the corresponding networks in the electric model. Models of indirect similarity include analog computers, investigative experiments in criminology, historical parallels, the life of separated twins, and experimental animals in medicine.

However, you should carefully use analogies because, in addition to the same patterns, different phenomena have different features as well. Therefore, not all conclusions about the model can be transferred to the original, and not all features of the original are present in the analog model. Sometimes the concept of "degree of analogy" is introduced associated with the extent of "overlap" of the compared theories.

The third type of real models is based on similarity, which is neither direct nor indirect. For example, letters — sound models; money — value models; various signs, signals, symbols, maps, and drawings contain relevant information. Compliance of such models with their originals is a result of agreement between their users, Let's call such models *conventional similarity models*.

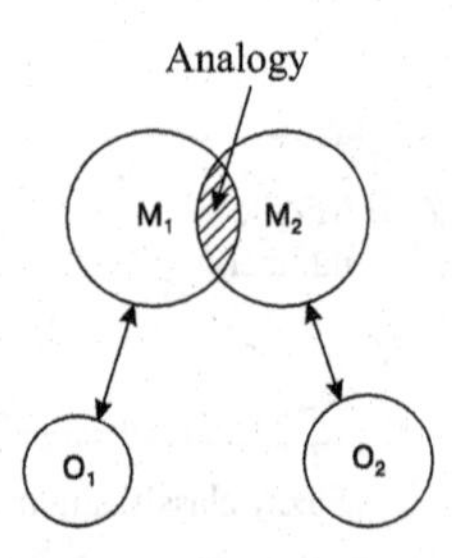

FIGURE 3.12 Analogy as basis for conventional similarity of models.

They work successfully but only as long as the agreements on their meaning (monetary reform, dead languages, secret signs, etc.) are known and respected.

Further analytical consideration of the set of all real models cannot be brought to define common elements: the variety of subjects used as models is too large.

It is possible, of course, to identify the elements of a specific real model (e.g., a geographical map), but the conclusions will be of a particular nature.

3.8 SYNTHETIC APPROACH TO THE CONCEPT OF A MODEL

In accordance with the synthetic method, explaining the nature of models begins with defining a metasystem of which the model is a part. It is possible to begin the selection of the metasystem from the above definition of the model as the image of the original. This definition has already identified two elements of the metasystem: the model and the simulated original.

An important feature of the model is that it is never identical to the original (even when they are trying to achieve it — counterfeit banknotes, copies of works of art, and other fakes). Often this is simply not necessary: each model is needed for a specific purpose, which requires only some (far from all) information about the original.

The purposefulness of the models has a number of important consequences.

The first is that the purpose of modeling is determined by a certain subject, who, therefore, must be included as another element of the metasystem.

A variety of purposes leads to a plurality of different models for the same original. For example, we should not be puzzled by the existence of several different definitions of something, or by the different testimonies of witnesses to the same event. As an example of the multiplicity of models of one object, to describe the different relations between the subjects in the applied systems analysis, three types of ideologies are considered (see Chapter 1); and political scientist R. Epperson distinguished the following five types of a form of government in society:

Rule by no one: anarchy.
Rule by one man: a dictatorship or a monarchy.
Rule of the few: oligarchy.
Majority rule: democracy.
Rule of law: republic.

Models of the phenomenon can even contradict each other (e.g., the corpuscular and wave theory of light). Models can be distinguished by type of purposes. For example, it is useful to classify models into *cognitive* and *pragmatic* ones.

Cognitive models obtain information about the outside world, represent the obtained knowledge, and are subject to changes when new knowledge is added to them.

Cognitive models (Figure 3.13) do not pretend to finality or completeness: there is always something unknown. In cognitive practice, it is customary to tolerate differing and even contradictory opinions. Scientific models are constantly questioned and checked for accuracy, and are continuously refined and developed.

Continuing to consider the relationship between the model and the original, we will focus on the content of information in the model. The original and the model are

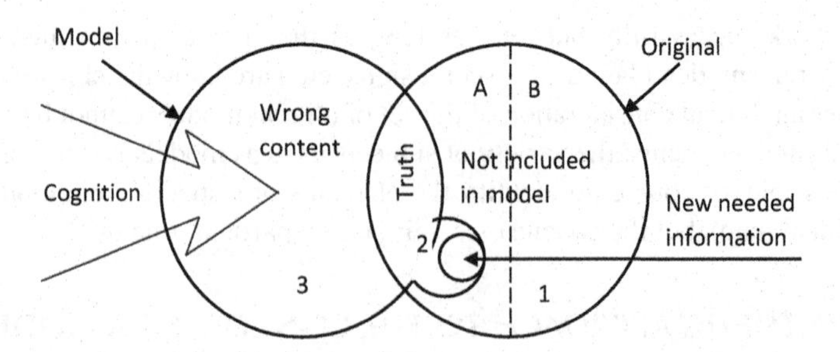

FIGURE 3.13 Scheme of cognitive modeling.

different things. There are a lot of things in the original that are not included in the model for two reasons: first, not everything that is known about the original will need to be included in the model designed to achieve a specific goal (zone A in Figure 3.13 depicts the known but unnecessary, including the wrongly considered unnecessary and not included in the model); second, there is always something unknown in the original, which cannot be included in the model (zone B in Figure 3.13). Zone 2 in the figure shows information about the original included in the model. This is true information, something common in the model and the original, thanks to what the model can serve as its (partial, specific) substitute or a representative. Let's pay attention to zone 3. It reflects the fact that the model always has its own properties that have nothing to do with the original, that is, *false* content. It is important to emphasize that this applies to any model, no matter how hard the model creator tries to include only true information in the model.

For example, the analytical function of time as a signal model reflects the fact that the signal is a certain time process. However, this model does not reflect the fact that the repeated signal does not carry new information as the first time. This model does not have the property of real signals to simultaneously occupy a finite time interval and a finite frequency band. In many (and if look closely, in all) theories, the feature of the model to contain false information manifests itself in the form of so-called paradoxes. For example, in the theories of electrostatics and gravity, infinity paradoxes occur at zero distances between interacting particles.

Pragmatic models transform reality in accordance with the objectives of the subject. They reflect the currently nonexistent, but desired (projects, plans, programs, algorithms, rules of law, etc.), and have a normative, directive character. This gives them the status of "the only true", which is clearly expressed in ideologies, religions, morals, standards, technical drawings, technologies, etc. In contrast to cognitive models, "adjusted" to reality, in transformative activity, reality is "fitted" to the pragmatic model (Figure 3.14).

Let's conclude the consideration of the relationship between the original and the model by emphasizing the inherent inaccuracy and approximation of the model. Even those aspects of the original that are intentionally displayed are described with some accuracy and approximation. Sometimes the approximation is forced by nature (lack of knowledge), and sometimes it is introduced deliberately for the simplicity of work with the model (e.g., linearization of nonlinear relations between variables).

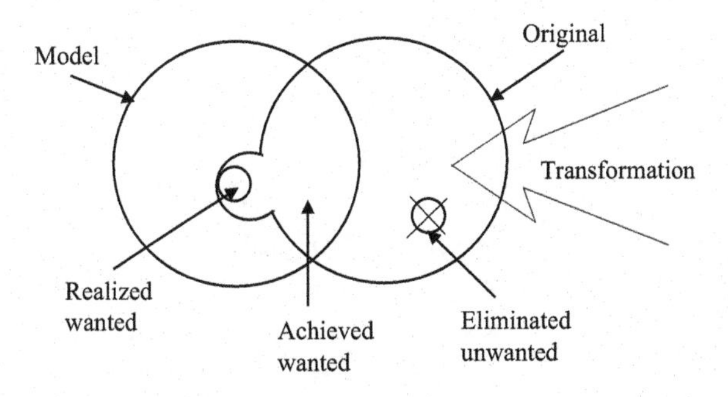

FIGURE 3.14 Scheme of pragmatic modeling.

3.9 THE CONCEPT OF ADEQUACY

At times, it is possible to achieve the same purpose with the help of various models (e.g., hiking using maps of different scales). It turns out that different models of the same object provide different levels of success in achieving the goal. This property of models is called the *extent of adequacy.* Usually, two levels of adequacy are used: the model successfully achieving the goal is called *adequate*, while the other is called an *inadequate* model.

Discussing the relation between such properties of models as adequacy and accuracy (truth) is of interest as they are not always compatible.

For cognitive models, whose destination is an accumulation of knowledge about the surrounding reality, they are synonyms. This is not the case with pragmatic models. Every one of us has told lies. Lie was preferred to truth because achieving the goal with the help of the lie was easier than with truth. Therefore, in certain circumstances, false models may be adequate (otherwise, lie would not be needed).

3.10 THE COHERENCE OF THE MODEL WITH THE CULTURE

You cannot read a book written in an unfamiliar language; it is impossible to listen to a record on a gramophone record without a gramophone; a fifth-year student would not understand the special course without the knowledge gained earlier. Similar examples illustrate the fact that for a model to realize its model function, it is not enough just to have the model itself. It is necessary that the model is compatible and consistent with the environment, which for the model is the culture (world of models) of the user.

When considering the properties of systems, this condition is called inherence: *the inherence of a model to culture of the subject* is a necessary requirement for successful modeling. The degree of the model's inherence may change, that is, increase (user training, the appearance of an adapter such as a Rosetta stone, etc.) or decrease (forgetting, degradation of culture) due to a change in the environment or the model itself.

Thus, another element should be included in the modeling metasystem — *the culture of the subject.*

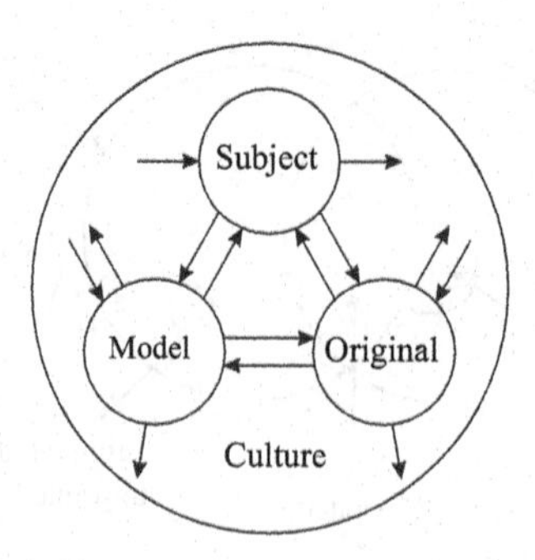

FIGURE 3.15 General scheme of modeling as a metasystem.

As a result, the scheme of the metasystem "modeling" can be represented as in Figure 3.15. In accordance with the synthesis methodology, to explain what a model is, it is necessary to discuss the relationship of the model with the remaining components of the modeling metasystem. This was the subject of the above reasoning. At the same time, we identified only those connections that are essential for the subsequent presentation of the technology of applied systems analysis.

The remaining links between the elements of the metasystem can be the subject of special consideration and are considered by various sciences.

3.11 HIERARCHY OF MODELS

Therefore, any activity of the subject is based on the use of models, that is, knowledge of what the subject is dealing with, and why he is performing the activity. In this case, models can describe both the real and the desired states of the system under consideration with varying degrees of detail. In this regard, it is useful to distinguish the levels of elaboration of information with which you have to deal. R. Ackoff proposed [1] the following classification (note the specific use of known terms):

Data — (what?) — description of the results of measurements and observations; experimental protocols; the original "raw" data.
Information — (composition?) — the result of a primary data processing; their ordering, classification, and grouping.
Knowledge — (structure?) — the result of secondary data processing; identifying links and patterns between groups and classes of data.
Understanding — (why?) — an explanation of the identified patterns; the construction of theories that give such an explanation.
Wisdom — (what for?) — information about why all this is necessary; is it good, should it be continued or stopped? that is, approach in terms of aesthetics, ethics, and ideologies.

In an attempt to emphasize the different significance of these levels of information, according to R. Ackoff, an ounce of this level is equal to the value of a pound of the previous one. It is possible to argue about quantitative relationships, but the qualitative difference is obvious. It is worth noting that in the existing education system, attention and training time is given to levels of information in inverse proportion to their actual importance.

Both above-discussed activities of a subject (cognitive and transformational) serve the external needs of interactions of the subject with the environment. However, there are other types of human behavior, neither cognitive nor practical, that serve the internal needs of a subject — arts and the eternal rest (especially, sleep dreams). They are also supported by modeling, but it's a subject of another book.

Questions and Tasks

1. Show that the cognitive and transformative activities of the subject are impossible without modeling.
2. Describe the analysis algorithm and list which models it generates.
3. Describe the synthesis algorithm and indicate which models it generates. Which one of them directly describes the object (phenomenon) under investigation?
4. What is an "abstract model"? In addition to language, what other examples of abstract models can you give?
5. What caused the diversity of languages?
6. Which is the simplest abstract model of the diversity of the reality surrounding us?
7. What is the difference between artificial and natural classifications?
8. What is a "real model"? Give three types of real models (classification by origin of similarity of the model to the original).
9. What is the difference between the use of cognitive and pragmatic models?
10. Why any model, besides the true content, has and (necessarily and inevitably) untrue content?
11. Which quality of a model is called adequacy?
12. What is the environment for the model?
13. Define the following terms:
    - model;
    - analysis;
    - synthesis;
    - abstract model;
    - language model;
    - real model;
    - classification (artificial and natural);
    - cognitive models;
    - pragmatic models;
    - the adequacy of the model;
    - culture (of subject, organization, and nation of any social system).

4 Control

4.1 ANALYTICAL APPROACH: FIVE COMPONENTS OF CONTROL

The first control component is the controlled object itself, *the system being controlled.*

Denote the outputs of a certain system S by the symbol $Y(t)$, and divide the inputs into externally controlled $U(t)$ and uncontrolled, but observable $V(t)$ (Figure 4.1). We know that there are inputs, unknown to us, but the unknown cannot be included in the model except through the notion of stochastic events: the observed values turn out to be random. Even so, the unknown is not reflected by the fortuity of the known.

The allocation of controlled inputs means that we consider the system S as a controlled object. The outputs $Y(t)$ are the result of the transformation of the system S inputs.

$V(t)$ and $U(t)$: $Y(t) = S(V(t), U(t))$, which allows you to act on $Y(t)$ by selecting different controls $U(t)$ (note that the symbols given are valid only for noninertial systems, although in reality the output of the system depends not only on the inputs at a given time but also on their history. However, for our purposes this is not yet significant).

The second mandatory component of the control system is *the control purpose.* In Chapter 2, we discussed in detail the concept of goal when considering the problem of the feasibility of all systems. We only recall that our goal concept includes not only the final desired state of the system (T^*, Y^*) but the entire desired path to it, $Y^*(t)$ (Figure 4.2). Recall also that no matter how hard we try to take into account all the limitations when formulating a purpose, it remains subjective: first, we took into account only what we know, and our knowledge is always limited; second, how exactly and how correctly we did it, that is, the result of our work, which inevitably bears the imprint of a person. So, the question of the actual attainability of the goal with the help of the S system remains open until the very beginning of the control process.

The control action $U(t)$ is the third control component. The fact that the inputs and outputs of the system are interconnected by some relation $Y(t) = S[V(t), U(t)]$ allows us to hope that there exists such a control action $U^*(t)$, in which output implements the goal $Y^*(t)$:

$$Y^*(t) = S\left[V(t), U^*(t)\right]. \tag{4.1}$$

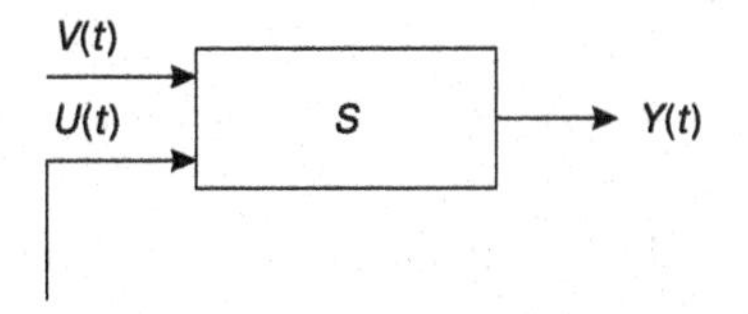

FIGURE 4.1 A general scheme of a system in environment.

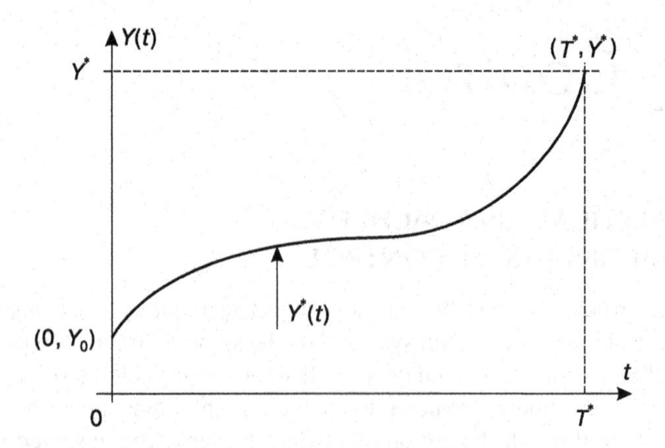

FIGURE 4.2 The purpose is the entire path from problematic to ultimate goal states.

But how do you know if it really exists, and if so, what is it?

To do this, you need to solve Equation (4.1) with respect to $U^*(t)$.

In this equation, $Y^*(t)$ (given) and $V(t)$ (observable) are known, but the operator S is usually unknown, which makes the problem unsolvable.

The solution is still to be found, and this leads to two types of control.

The first is to apply to the controlled input some kind of influence $Ui(t)$ and see what happens. If the output is the goal $Y^*(t)$, we are very lucky. If not, submit another impact $Uj(t)$ and observe the result. To continue in this manner until the desired result is achieved, that is, look for the desired input $U^*(t)$ by sorting the effects on the system S itself.

Though sometimes this method is the only possible way (e.g., finding a way out of the maze), more often such a control method is unreasonable for several reasons. For example, the set of possible $U(t)$ can be so large (and even infinite) that it is unrealistic to hope for an occasional successful hit. Another important reason is high losses in case of an incorrect decision. For example, S is a school, $U(t)$ is a teaching method, and $Y(t)$ is school graduates. It is clear that the search method is inappropriate here. Therefore, the search for the desired control action on the controlled object itself is often unreasonable and unacceptable.

The second approach is based on the use of all available information about the managed object. This implies that the search for the desired control should be carried out not on the system itself but on its model.

Thus, *the system model* becomes the fourth component of the control process. Instead of solving Equation (4.1), we now have to solve Equation (4.2) with respect to the control action $Um^*(t)$:

$$Y^*(t) = Sm\left[V(t),\ Um^*(t)\right] \tag{4.2}$$

where $Y^*(t)$, $V(t)$ and Sm are known, that is, the model of the system. In principle (we will leave aside technical difficulties), such an equation can be solved. This will be rational and prudent control.

Of course, the search for control on the model also requires losses (the cost of the modeling process), but these losses are incomparably less than those we would have incurred looking for the right control on the system itself.

All actions necessary for successful control must be performed. This function is usually assigned to a specially created system for this (the fifth component of the control process), called the *control unit or the control system (subsystem), control device*, etc.

In reality, the control unit may be a subsystem of the controlled system (as the plant management is part of the plant, the autopilot is part of the aircraft), but it may also be an external system (as a ministry for the departmental enterprise, as an airport controller for landing an aircraft). There are complexities involved in constructing a model of composition, as discussed in Chapter 2.

So, the control system may look like it is presented in Figure 4.3, which shows all five components of the process of purposeful influence on a system.

We have now successfully established the first two mandatory steps of the control process (the model of structure of control):

1. find the required control action $Um^*(t)$ on the model of the system and
2. execute this effect on the system.

4.2 STAGE OF FINDING THE DESIRED CONTROL ACTION

How to use the Sm model to find the best control action? Having used the evaluative word "best", we must indicate precisely in what sense this evaluation is used, that is, to set quality criteria. It is clear that the control is "better" when the output of the system $Y(t)$ is "closer" to the purpose $Y^*(t)$. But we'll look for this control on the model, so at the control search stage, we'll have to consider as the best $Um^*(t)$, which will maximally bring the output $Ym(t)$ of the model to the purpose $Y^*(t)$.

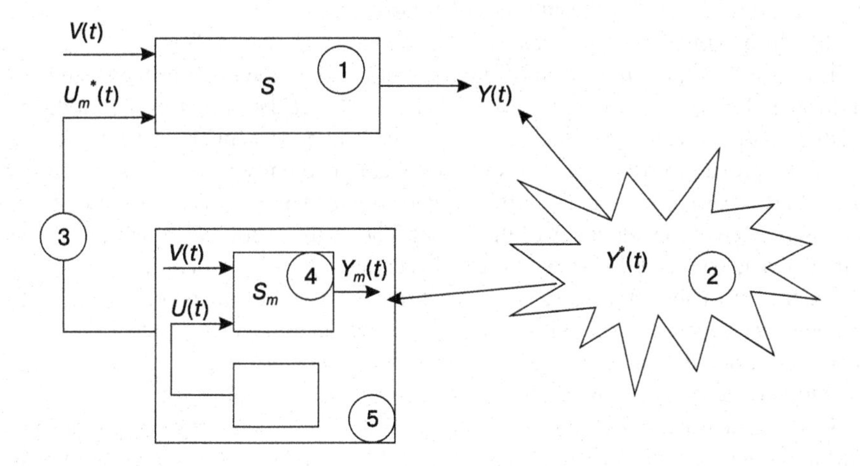

FIGURE 4.3 General scheme of the control system.

If the outputs $Ym(t)$ are measurable numerically, then some numerical criterion ("distance" between two functions) $r = r(Y^*(t), Ym(t))$ is introduced, which would be zero if the compared functions coincide and increase for any of their differences. Such "distances" can be entered in many different ways. For example:

$$r_1 = max\left|Y_m(t) - Y^*(t)\right| \cdot r_2 = \int_{t_1}^{t_2} \left[Y_m(t) - Y^*(t)\right]^2 dt$$

Having chosen some measure of difference between the two functions, we have to solve the problem of finding such $Um^*(t)$, which delivers the functional r minimum (better, zero):

$$U_m^*(t) = \left\{ \underset{\{U(t)\}}{minr} \left[Y^*(t), Y_m\left(V(t), U(t), U^*(t)\right) \right] \right\} \tag{4.3}$$

For the purposes specified non-numerically (in qualitative terms), methods of fuzzy logic still allow to introduce some evaluation of the proximity of the result to the purpose.

4.3 SYNTHETIC APPROACH TO CONTROL: SEVEN TYPES OF CONTROL

After applying the influence $Um^*(t)$ to the controlled input of the system, the system will produce some output process $Y(t)$:

$$Y(t) = S\left[V(t), Um^*(t)\right],$$

which is the transformation of the inputs by the operator S of the system. In this case, there may be different outcomes that require different actions to manage the system. This generates several different types of control.

The first type is *simple system control or programed control.*

Let's start with the most desirable case, that is, when the input of the system S is influenced by $U_m^*(t)$, which provides the target $Y^*(t)$ at the output of the model S_m, leading to the same result at the output of the controlled system. This means that our S_m model proved to be *adequate* as the S system obediently fulfilled the given purpose. In this case, the system S will be called *simple.* The simplicity of the system is a consequence of the adequacy of the model. The control action $U_m^*(t)$ in this case is called a *program*, and this type of control is called a *programed control.*

Such a most favorable case can sometimes be realized in practice. Examples include serviceable household appliances, various automata, computers, weapons, an effective and dependable employee, an ideal soldier, etc.

The second type of control is *the management of a complex system.*

Consider another extreme case, when the system S does not respond in the same way as the model to the control action $Um^*(t)$ found on the model: $Y(t)$ does not coincide with $Y^*(t)$. We denote this situation with the appropriate terminology.

Let's start with the fact that our model did not allow us to achieve the goal; our S_m model is *inadequate*. System S behaves in an unexpected way and does not submit to our control ("this damn thing behaves not in the way it should be!"). We will call such a system *complex*. The reason for the complexity of the system with this approach is the inadequacy of its model Sm.

We emphasize that we have introduced a special definition of complexity. There are several other definitions, some of which are associated with the concept of complexity with multiplicity and variety of the components, while others with the multidimensionality of control components and, especially, with the law of universal interconnection and interaction in nature. We will use the term "complex" only in the sense of insufficient information about the controlled object. From this viewpoint, complexity is not a property of the system, but a property of those who work with the system.

Obviously, the management of a complex system is reduced to obtaining the missing information about the system and the subsequent use of this information for the next control act. This means that we must improve the model of the system and enhance its adequacy.

We will proceed from the assumption that in building the S_m model we used all the available information about the system from textbooks, monographs, reference books, Internet, and experts. Thus, the only source of information remains the system itself, and the only way to extract this information is to experiment with the system.

Experiment is a question to the system, to which it gives an honest answer. We have already asked one question. By applying to the controlled input the impact $Um^*(t)$, we asked the system: "Honey, will you give out the impact on the output $Y^*(t)$"? And she replied: "No, I'm not like that! I respond with the function $Y(t)$". This received information should be included in the model by transmitting information through the feedback circuit, as well as changing and correcting the model so that it responds to $Um^*(t)$ with the same function $Y(t)$ as the system (Figure 4.4). Now the Sm model has become more similar to the S system, at least in this example.

We use the new, revised, and augmented model S_{mi} to search for the next control action (here we introduce in the diagram the index "i" of the next step $i = 1, 2, \ldots$), $U_{mi}^*(t)$. Such steps are repeated, which gradually improve the model and increase its adequacy.

Therefore, the control algorithm of a complex system is as follows:

1. At this i-th point in time, models Smi of system S are searched by some method (and methods can be different: random search, gradient descent, sorting out, etc.) and control action $Umi^*(t)$, which provides the target $Y^*(t)$ functions at the output of this model.
2. Impact $U_{mi}^*(t)$ is fed to the controlled input of system S.
3. The output of the system $Yi(t)$ is observed and recorded.
4. If $Yi(t)$ and $Y^*(t)$ differ, the model is corrected (due to its varying parameters) so that the corrected model $S_{m(i+1)}$ is repeated as accurately as possible at its output, giving the system response $Yi(t)$.
5. Return to point 1 ($i \rightarrow i + 1$).

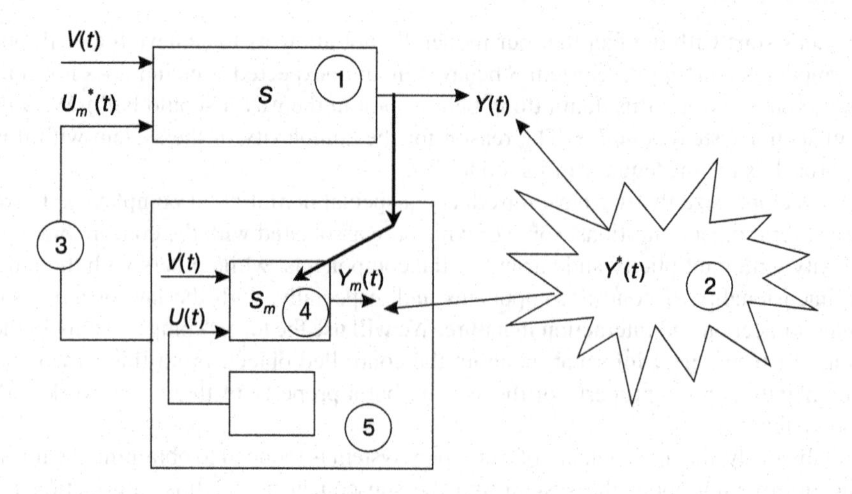

FIGURE 4.4 Scheme of controlling a complex system.

Let us again discuss the features of the control algorithm of a complex system.

First, the algorithm has a cyclical and repetitive character. With each cycle S_m improves and becomes more adequate, which increases management efficiency and reduces the complexity of the system. In some cases, a complex system can be turned into a simple one in a finite number of steps. An example is the case when you forgot the cipher you dialed from an automatic camera. An improvement in the model consists in replacing its string "with number X may open" after unsuccessful sampling No. X with the string "with number X not opening", and in a corresponding reduction in the number of remaining variants.

In other cases, the model is corrected by changing its parameters. For example, if a model is an equation, its coefficients change, indicators and members of the equation are added or eliminated, etc. If the model is a physical device, its settings, adjustments, switching, etc. also change. Sometimes these actions lead to a sufficient adequacy of the model, that is, they simplify the system.

However, there are systems whose complexity mankind cannot exhaust despite all efforts (nature, society, economy, thinking, etc.). This does not mean that their study is in vain, it is simply infinite. They are sometimes called *very complex* systems.

Second, since at each step "not quite the goal of $Y^*(t)$" will be obtained, we will suffer losses. Such is the price of ignorance. We can only minimize the inevitable losses in the management of a complex system.

This can be done only completely and without any loss using the information obtained in the next experiment (control step), that is, to make the adjusted model as accurately as possible to simulate the behavior of the system at each of the previous steps.

Now it is time to give a widely used name for this method, although with some reluctance because of linguistic peculiarities. While developing a professional terminology for the needs of the theory and practice of governance, each next control action was called a *trial action* or just a sample, and the discrepancy between $Yi(t)$ and $Y^*(t)$ an *error*. The control algorithm of a complex system is called the trial and error method. Because of this terminology, some confuse it with "poke method". The fundamental difference between them is that the desired effect is not sought on the system itself (in the "poke method"), but on the model of the system which is corrected in the course of control. We can say that the "poke method" is the worst method of managing a complex system, while "trial and error" is the best. But even in this case the loss is inevitable: for the ignorance you have to pay.

It is important to emphasize that even with the obvious infinity of knowledge of the very complex systems, progress is still possible by trial and error method; although the exact task of the final goal will have to be abandoned, it is possible to go in its direction, overcoming specific "today's" obstacles and defining specific restrictions, within which there is freedom for trial and error.

The third type of control is *parametric control or regulation.* Consider the intermediate case between the first two.

By submitting $Um^*(t)$, we can observe that at first the system follows the desired trajectory $Y^*(t)$, but after a while there is a discrepancy between $Y(t)$ and $Y^*(t)$ (Figure 4.5). Of course, this means that something is wrong with the model.

However, it often turns out that amending the model of the system is impractical. For example, a plane driven by an autopilot goes off course by a gust of wind. It makes no sense to introduce this impulse even into the model of the environment $V(t)$ as it will not happen again. But we need to return the system on desired trajectory $Y^*(t)$. *Now* it is possible by making changes to the system itself.

The first possibility is to change the *system parameters* without changing its structure. The parameter is changed so that the system returns to the target trajectory, and at further deviations of $Y(t)$ from $Y^*(t)$ this action is repeated. A good example is driving a car. The target trajectory is the right side of the road (in England, Japan, and other countries on the left). The driver changes the position of the steering wheel (and other adjustable parameters of the machine), holding the car on the target trajectory (Figure 4.6).

To implement this type of control, you must perform the following functions:

1. keep the reference trajectory $Y^*(t)$ in memory;
2. follow the real trajectory $Y(t)$;

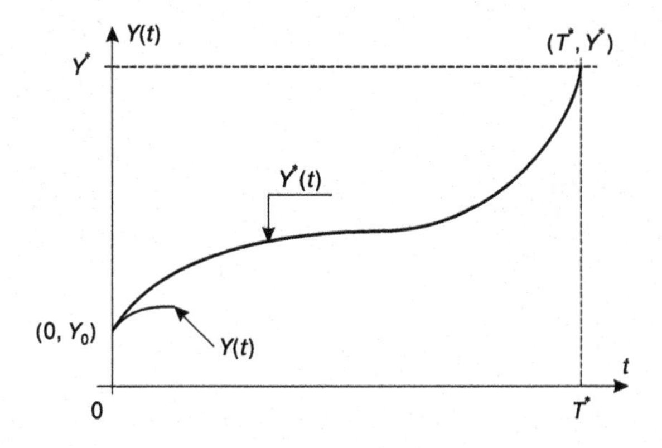

FIGURE 4.5 The first stage of regulation.

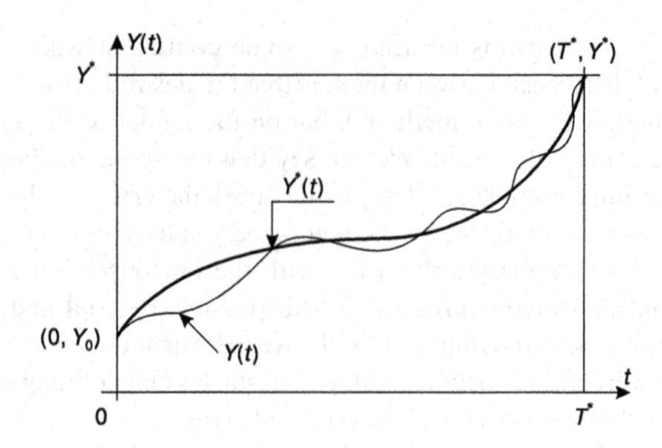

FIGURE 4.6 The full cycle of regulation.

3. detect the current difference between $Y(t)$ and $Y^*(t)$;
4. develop, determine, and calculate the corrective additional to $Um^*(t)$, as well as the impact on the parameter(s) of the system;
5. perform this action on the system, returning it to the reference trajectory.

To perform these functions, you must create a special device, an additional system. This device is called *the regulator*, and the method of control is called *regulation.* The control scheme now looks different (see Figure 4.7, and compare it with Figure 4.3). The regulator is represented by a square "R". It is placed on the scheme at a position that allows it to be attributed either to the system itself, to the system controlling, or to be considered as an independent system. It is important to perform these functions, and where they will be physically performed depends on the circumstances. For example, the control of the aircraft on different parts of its trajectory can be performed by an autopilot, a pilot, or an aerodrome dispatcher.

If the software control is called control without feedback, the regulation control is a *negative feedback* (emphasizing the desire to reduce the deviation from

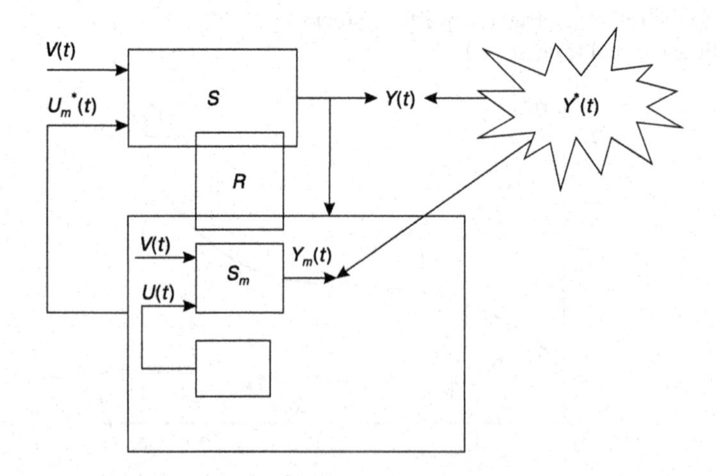

FIGURE 4.7 Scheme of a regulated system.

the reference path; with positive feedback, the deviation tends to increase, as, for example, in oscillators).

To implement the regulation, it is necessary that the system deviates from the reference trajectory not too quickly and not too far, so that, by changing the parameters, it can be returned to the target trajectory $Y^*(t)$. In the mathematical theory of regulation, this condition is called "small" deviations.

Unfortunately, this condition is sometimes not satisfied. For example, the car can go to the icy section of the road, and no manipulation of the wheel and brakes will be able to keep it on the road. We find ourselves in an environment where a different type of control is required.

The fourth type of control is *control by means of changing structure.* When the system deviates so quickly and so far from the target trajectory that it cannot be returned to it by changing the parameters, we have two options: pessimistic and optimistic. Pessimistic means humility before the impossibility of achieving the ultimate goal, and sometimes even death. Optimistic is associated with recognizing the facts and making an attempt to achieve (T^*, Y^*) by some other means. The fact is that this goal is unattainable for the *existing* system. But maybe it is achievable for *another* system?

Change the structure of the system at the moment Ts (Figure 4.8), thus creating a new system, with the hope to arrive at the point (T^*, Y^*) even if on a different trajectory $Ys^*(t)$. This control is called the control by means of changing the structure. You can distinguish between cases where the new structure is created only from parts (perhaps not all) of the old system, and cases of involvement in the structure of new elements from the outside.

Various options meet the multiplicity of names for this type of control: reorganization, modernization, restructuring, self-organization, etc. As examples, we can cite surgery, change the administrative management scheme, ballast discharge from the balloon, intake or blowing water from submarine tanks, extension to the building, prosthetics, etc.

It is clear that we can meet the case when no combination of available elements ensures the achievement of the ultimate goal. This means it is impossible and inappropriate to control by structure, the potential of which is exhausted.

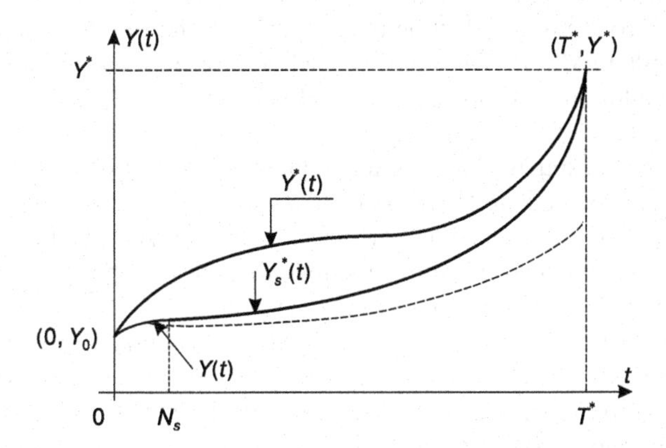

FIGURE 4.8 Trajectories of system under changing its structure.

The fifth type of control is *control by purposes*. The output is again seen not in hopeless lowering arms, and to recognize the fact and make an optimistic conclusion. The fact is that no use of the limited means and resources available can realize the desired state: this purpose in these conditions is unattainable.

It remains to change the purpose, lowering the level of claims, and to refocus on achievable deadlines and (or) other parameters of the final state. This is the fifth way of control, that is, control by purposes.

It is possible to distinguish the purposes that are unattainable in principle. The detection of the unattainability of some of these goals is a reason to abandon the desire for them.

There are, however, other obviously unattainable goals, but attractive and worthy aspirations for them, and most importantly, by allowing unlimited approach to them; such goals are called *ideals* (harmoniously developed personality, absolute sports superiority, knowledge of the truth, absolute good, etc.), and people spend a lot of energy to strive for these goals.

There are goals that are unattainable in some conditions, but achievable in others; there are goals, the achievability of which is desirable, but not proven, although not refuted (artificial thinking, antigravity). There are also goals that are achievable but not achieved because of ineffective or erroneous governance. However, in some cases, determining which of the options we are faced with is not easy.

In the practice of governance by objectives this is quite common, especially in the administration and management: rotating personnel, offering feasible jobs, retraining, and other methods of personnel management. It should only be borne in mind that the change of purpose for any individual is a painful process, the more difficult the higher the level of the purpose has to be changed (awareness of the unattainability of the goal can sometimes even make the subject's life further meaningless). So, this method requires caution.

The sixth type of control *is the control of large systems*. The first two types are based on improving the system model, the third and fourth on changing the system itself, and the fifth on changing the purpose. There is another factor that affects the quality of control and requires a new way of doing it. This is the *timeliness* of the control action. The statements "The train has already left", "Where were you yesterday?", "Strong back mind", "Wit on the staircase", "The ship has sailed", "No use crying over spilt milk", "It's easy to be wise after the event", and "You've missed the boat" display a folklore fact of the futility of belated decision, even the best one in all other respects.

The delay in choosing the best possible solution is caused by the fact that to evaluate each of them you need to "play" it on the model of the system, and this requires some time. However, the time allowed to develop a solution may be limited: after this time, management becomes meaningless. It is necessary to operate in real time and to model management in an accelerated manner.

It may turn out that the time required to find the optimal solution exceeds the maximum allowable for executing the control intervention. Then the very opportunity to find the optimal solution becomes superfluous. This is necessary to operate and requires developing another method of management.

We will now frame a situation with which we are faced with special terms. The system for finding the optimal impact on which there is enough information resource (the model is adequate), but not enough time to sort out all of it, we will call a *large* system, otherwise a *small* system.

For example, we can cite the Soviet economy when the intersectoral balance was brought with a delay of three to four years. (In the centralized economy of a huge country, it was necessary each year to maintain a balance between millions of produced and consumed products.) It was believed that this is one of the main reasons for the low efficiency of economic management in the country.

Another example was given at the time by the Novosibirsk Academy computer center, which implemented a very developed multicomponent model for accurate weather prediction for the day ahead, but the power of the then computer center made it possible to obtain a forecast only in a few days.

It is clear that the reason that the system turns out to be large is not the magnitude itself, that is the bulkiness of the system, but the insufficient speed of search and comparison on the model of control options, that is, the lack of time.

Therefore, the first and the most effective way to control a large system is to turn it into a small one, speeding up the modeling process. For example, replacing the modeling computer with a faster one, parallelizing the optimization algorithm, delegating parts of manager's powers to assistants, etc.

However, this method can encounter insurmountable difficulties (e.g., there are no more powerful machines, there were no suitable personnel, there is not enough finance). Therefore, in real practice, timely decision-making gives the result. Not the best, but a timely decision is better than no or belated decisions. We need to abandon the expectations of obtaining the optimum variant and to receive the first of the resulting satisfactory. Often to obtain a weak but fast solution we turn to various simplifications of the model (dimension reduction, linearization and other simplifying approximations, rounding of exact numbers, etc.). This accelerates tackling of a difficult situation for the manager, acting under the time constraints._Sometimes, however, behind this lies the inability to work better.

The seventh type of control is *control under unknown ultimate goal.* In addition to the first type of control, when everything necessary for the implementation of the goal is available, the rest of the control types are associated with overcoming the factors that prevent achieving the goal: lack of information about the control object (the second type), small difference between real environment $V(t)$ and its model $V_m(t)$, slightly deviating system S from the target trajectory (the third type), the discrepancy between the emergent properties of the system and the goal (the fourth type), the lack of material resources, making the goal unattainable and requiring its replacement (the fifth type), and the lack of time to find the best solution (the sixth type). All these cases are based on the assumption that the end is known.

However, in real life, there is another situation: when it is necessary to control current events, but the ultimate purpose is *incomprehensible* and *unknown.*

How to govern if you are deprived of the opportunity to specify the ultimate goal? In this case, we go to the seventh type of management — *control in the absence of information about the ultimate goal.*

From the definition of the purpose adopted by us in Chapter 2, it is logical that the uncertainty of the final goal is followed by the uncertainty and the trajectory of movement to it. However, the control actions for any type of control are aimed at moving along this path with the maximum achievable proximity to it. This aspiration can be realized in at least two ways in this situation.

The first is to give a subjective, a priori definition of the ultimate goal, and then act on the previous schemes.

A clear (but not the only) example of this gives us the governance of large social systems. What is the meaning of life? What is the purpose of social development? Ideologies give ready answers to this. However, these answers are only hypotheses.

Different communities adhere to different ideologies by subjectively giving preference to one or another ideal. History has already shown the lifelessness of some of them (slave and feudal system), has revealed the acute shortcomings of others (tyrannical, dictatorial regimes), and has utopian character of thirds. We are witnessing the decline of society toward the ideals of liberal democracy. But even in democratic ideology, some fundamental goals are contradictory. For example, the ideas of equality and freedom are incompatible: freedom is impossible with equality, and equality is impossible with freedom. The attempt of the French revolution to unite them with the help of "brotherhood" looks naive or in any case unconstructive. The applied systems analysis offers in this case one more ideal — equality and equivalence of each individual and improving intervention as a way of realization of this ideal (see Chapter 1).

An interesting option for implementing the democratic ideal (majority decision-making) is a two-party system. One party gives priority to freedom, while the other to equality. Both ideals are attractive, but incompatible. Society elects "socialists", that is, begins to realize equality (in particular, carries out nationalization of large branches of economy and realizes social projects). However, the process of equalization inevitably constrains the initiative of the subjects, and the development of society slows down. When this becomes obvious and undesirable, society elects "liberals" who start privatization and unleash personal initiative through freedom and private property. This accelerates the development of the economy, but increases inequality between the rich and poor segments of society. This causes an increase in tension in society, as well as in the sense of "injustice". Here and there is an opportunity to elect a party to power, preaching equality. A concrete example is England with its Labor and Tory parties.

The fact that this is a manifestation of some more general system regularity gives an example of optimizing the smoothness of the trajectory of the TU-154 heavy transport aircraft, as studied by Professor Y. I. Paraev of Tomsk State University. Aircraft flaps can be put in any position between the two extremes. It turned out (this is a strict mathematical result!) that the optimal control of such an inertial system as an aircraft consists in switching the flaps from one extreme position to another, that is, at correctly selected points in time. No intermediate position! There is a clear analogy with the two-party system. Maybe that's why the world is drifting multi-party systems in the direction of bipartisanship?

Yet, let us recognize that any social ideology that affirms its vision of the ultimate goal actually offers a hypothesis, the truth of which is a matter of faith in it and subsequent verification in practice.

There is, however, a different approach to management if the ultimate goal could not be clearly defined, but it is hoped that it exists. If this is the case, then there must be a trajectory to it. It is also unknown, but you can try to explore the nearest neighborhood around the current state and determine the most preferred direction of the next step within that neighborhood. Then take that step and act in the future as well.

This method is actually implemented in various areas.

In biology, this is called evolution and natural selection. In management theory (understood broadly), it is called incrementalism (making small but necessarily improving changes). In the mathematical theory of optimization, several ways of finding the extremum of the function of several variables (coordinate steps such as the Gauss–Seidel method, random search, the method of the steepest descent along the gradient, etc.) are proposed. In social systems, we have already mentioned Kautsky criticized by Marxists with his slogan "the goal is nothing, the movement to it is everything".

Of course, success is not guaranteed on this path. The objective function, the existence of which we are sure, can have several extremums, and we can get not in the global, but in the local extremum (e.g., dead-end branches of the evolution of living populations; the choice of satisfactory, not optimal solutions in the management of social systems; stuck in local extremums in mathematical optimization).

One gets the impression that the only absolutely universal purpose of the existence of any system is its very existence ("the meaning of life is in itself"). Only after surviving, the system can begin to pursue some (any) other goals.

4.4 SUMMARY

Let's summarize the results.

First, we considered the composition and structure of the control process, highlighting its five components (the object of management, the purpose of management, the control action, the model of the system, the subject of management) and discussing the interactions between them.

Second, we noted that after performing two initial operations (finding the desired control action on the system model and its execution on the system), the reaction of the system may be different. This requires specific management actions in each case, which allowed us to identify seven situations with special types of management behavior: program management, trial and error method, regulation, management by structure, management by goals, management in time deficit, and management in the absence of information about the ultimate goal. Each of them describes in detail the control algorithms and the limits of their capabilities.

Special attention should be paid to the special definitions of "complex system" and "large system" that we have introduced. They are the pride of the Tomsk school of systems analysis because before that different authors used these terms ambiguously, as synonyms, then used only one of them, then gave them a partially overlapping meaning. Clear distinguishing of these terms solved a few problems.

The first — the indication of different causes of complex and large systems has allowed us to offer different specific ways to overcome the difficulties associated with them: to combat the complexity additional information is needed to turn a large system into a small, accelerating decision-making.

The second — these qualities of the system can be combined in all four possible variants, requiring different approaches. The fuzzy division of the concepts "complex" and "large" can lead to difficult situations, according to the academician V. M. Glushkov, who did a lot for the development of computer technology and its implementation in the governance of the national economy of the USSR. At all levels, he argued that the main reason for the already manifested inefficiency of total economic planning is the insufficient capacity of the fleet of computers. When the Prime Minister A. N. Kosygin invited him to submit an application for any type and number of computers to solve this problem, it became clear that this action alone cannot solve the problem since the Soviet economy was not only a large but also a complex system.

Additionally, the classification of control types given in this chapter cannot be absolute and universal (as, indeed, any classification). As a model, it simply describes the variety of real-world control options. In life, there may be cases when the management of one system is used simultaneously or alternately a combination of different types of control. On the other hand, as a model, this classification is targeted, and other classifications may be required for other purposes. For example, in some cases there are automatic, semiautomatic (automated), and manual controls; these types are used in the control of machines, aircraft, manned spacecraft, etc.

Another classification will be required to distinguish management from other types of control: it is not a single-purpose management, not the management of the technical system, not the administrative management, not the automatic control, etc.

In conclusion, we note that in the Russian language the word "управление" has a very broad meaning. It includes such concepts as administration, command, management, and control of a technical device (machine tool, car, weapon, missile, etc.). It is interesting to know that in English there are words for specific types of control (government, management, control, administration, guidance, driving, etc.), but the general term equivalent to Russian "управление" is absent (the closest but not identical is processing).

Questions and Tasks

1. Which five components ensure the implementation of the control process?
2. Under what conditions is the search for control action on the system itself unreasonable and unacceptable?
3. What is a simple system? What is the reason for simplicity?
4. Which system is called complex? What is the reason for the complexity?
5. Describe the algorithm of trial and error. What features does it have?
6. What distinguishes the method of trial and error from the "poke method"?
7. List the functions of a regulator.
8. What is management of goals? Under which conditions is this type of control applicable?
9. What is the "large system"? What are the options to manage it?
10. Give examples of systems that would be both small and simple, small and complex, large and simple, large and complex.

Part I References

1. R. L. Ackoff. *Re-Creating the Corporation: A Design of Organizations for the 21st Century.* New York: Oxford University Press, 1999.
2. N. Wiener. *The Human use of Human Beings: Cybernetics and Society.* Boston, MA: Houghton Mifflin, 1950; 2nd ed., 1954.
3. R. L. Ackoff, F. Emery. *On Purposeful Systems.* New York: Aldine, 1972.
4. J. Gharayedaghi. *Systems Thinking: Managing Chaos and Complexity.* Boston, MA: Butterworth- Heinemann, 1999.
5. P. B. Checkland. *Systems Thinking, Systems Practice.* Chichester: John Wiley & Sons, Ltd., 1999.
6. M. C. Jackson. *Systems Thinking: Creative Holism for Managers.* Chichester: John Wiley & Sons, Ltd., 2003.

Part II

Systems Practice: Technology of Applied Systems Analysis

5 Technology of Applied Systems Analysis

In accordance with our system concepts, set out in Part I, the transition from the state of a problem situation to the state of a desired final goal — the solution of a problem — must be carried out systematically and orderly by consistently performing certain steps. In addition, each stage has its own structure of smaller steps, which should be followed fairly strictly; its violation may adversely affect the quality of the result of one stage and, consequently, of the entire process. In addition, at each stage, there are dangers to make a mistake or fall into a "trap"; you need to know about the possibility of such failing actions and use suitable techniques to avoid them. The technology of applied system analysis is presenting this entire algorithm with a description of all the features of each stage.

Having come to the system analyst with a problem the client could not solve himself, he initiates the systems analysis procedure. The analyst will take up work to solve any problem, but only under certain conditions. These conditions are absolutely necessary (although not sufficient) for success. Without them, experienced analysts simply do not get to work. Let us list these conditions, although their full meaning and necessity will become clear later after getting acquainted with certain points and subtleties of the technology.

Success conditions for system research:

1. guaranteed access to any necessary information about the problem situation (while the analyst, for his part, guarantees confidentiality, in case of its necessity);
2. guaranteed personal participation of the first persons in organizations, obligatory participants of the problem situation (heads of problem-containing and problem-solving systems);
3. the rejection of the requirement to preliminarily formulate the necessary result ("technical task", "terms of reference") since there are many improvement interventions that are not known in advance, all the more chosen for implementation.

5.1 OPERATIONS OF SYSTEMS ANALYSIS

If the client agrees to the terms of the contract, the analyst proceeds to the first stage, after which he starts the second and so on until the last stage, at the end of which the implemented improving intervention should be obtained (Figure 5.1).

From the properties of the composition model, we know that the whole can be divided into parts differently (the third property of the systems); therefore,

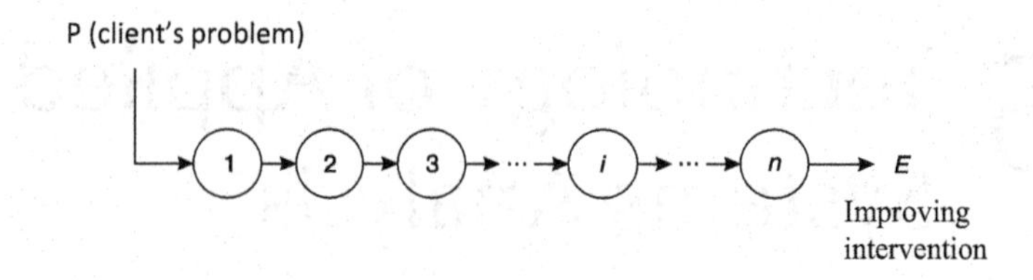

FIGURE 5.1 Technology of solving a simple problem.

different authors offer descriptions from different numbers of stages: descriptions can be enlarged and divided. For example, the path from posing a problem to solving it can be divided into two stages: research and resolution. Each of them can be divided into two stages: examination and diagnosis, planning and implementation of the plan, and so on, in full accordance with the method of analysis (see Chapter 3 in Part I).

We prefer to present a system analysis algorithm in a dozen stages. The scheme of presentation of each stage is standard: its input is given; it is described what and of what quality should be obtained at the output; it is indicated what actions and how should they be performed; it draws attention to possible difficulties, mistakes, and "traps" and gives advice on how to overcome them.

It would be wrong to think that a consistent linear passage of all stages will always lead to the desired solution. This happens only when the contractor knows in advance the complete solution to the client's problem. For example, when a teacher conducts classes with a student, teaching him how to add fractions, explaining to him every step in the process.

However, in the practice of applied systems analysis, the final result is not known to anyone in advance; it will be gradually formed during the analysis. Therefore, systems analysts refuse to accept from the customer "technical task" at the beginning of the work, which defines what should be the result of their work; therefore, systems analysts do not agree to "justify" someone who has already made decisions. A systems analyst cannot be compared with a doctor who diagnoses and prescribes treatment, and not even with a teacher who teaches what he knows, but rather with the head of the creative team who has taken on a new topic without knowing how it will end.

Consequently, the subsequent stages of the analysis may reveal incompleteness or error of some of the previous stages; it will be necessary to return to it and correct the discovered deficiency, once again going through the stages already passed. The scheme of the analysis procedure is shown in Figure 5.2.

The algorithm for solving a particular problem will not be linear, and it will contain cycles and returns. In fact, this is a different representation of the trial and error method: the solution to the problem is to overcome the complexity (!). The more complex is the problem, the more returns will be required (their number can be considered a measure of complexity of the situation).

However, in the future presentation of the technology, we will discuss the stages sequentially, one after another. The actual, real progress can take the shuttle path until you reach a final decision.

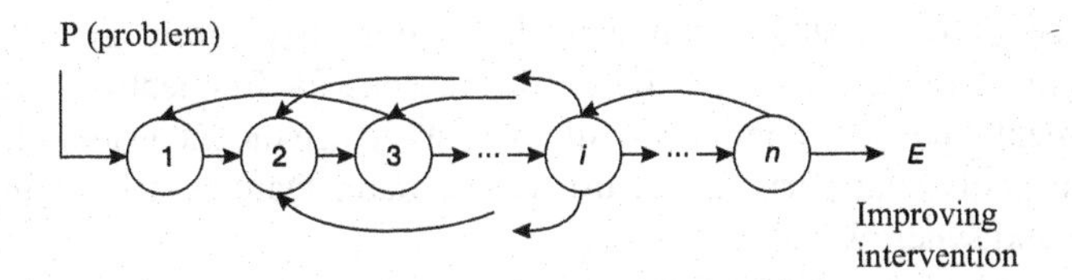

FIGURE 5.2 "Trials and errors" method for solving complex problems.

A common misconception, the main cause of failure, is the opinion that for any problem situation there is a simple and a correct solution. This is true only for cases where the set of problems is connected in a linear sequence, and the problem of the strength of the entire chain is to strengthen the weakest link.

However, such a situation is a rare exception; problems form a system whose emergence does not allow one problem to be solved individually and requires an effort to create an improving intervention (see Chapter 1, Part I).

Finally, it should be emphasized that both the operations of the systems analysis and the specific variant of the progressive-return process according to scheme of Figure 5.2 depend on the paradigms used (vision of the world, the set of ideas and assumptions on which we act) and metaphors (specific ideas about the system under consideration). Therefore, further presentation of the stages of the system analysis will be conducted taking into account different options for their execution.

5.2 ABOUT VARIOUS OPTIONS FOR SOLVING PROBLEMS

Even "hard" problems sometimes have various solutions (e.g., proofs of a mathematical theorem). For "soft" problems, this is a rule, not an exception. This is particularly evident and typical of the problems encountered in the management of social systems, with their diversity, complexity, and poorly predictable variability.

In response to the needs of managers in specific management practices in terms of the complexity of the managed system and the turbulence of the environment, there are various techniques advertised as a panacea for all managerial ills. Here are just a few of them:

- value-added analysis;
- core competencies;
- balance-sheet accounting;
- total quality management;
- continuous improvement;
- optimization of staffing (downsizing, rightsizing);
- user orientation (customer focus);
- planning in the form of scenarios (scenario planning);
- business process re-engineering;
- self-learning organizations;
- knowledge management;
- establishment of criteria and standards (benchmarking).

The practice of using such techniques shows that they rarely give the desired improvement. Although there can be several reasons for frequent failures, there is one fundamental panacea common to all: inability to see the integrity of the system, focus on its individual parts, disregard for the main features of systems, that is, emergence and synergy.

Modern applied systems analysis offers a holistic and creative approach, which focuses on two "system-forming" factors: (1) the integrity, emergence, synergy of the system (the inadmissibility of a separate consideration of any part when the goal is to improve the whole system), and (2) the entry of the system as part of a larger, embracing system, and the interconnectedness of the system with other systems in the environment (the need to take into account the integrity of the encompassing metasystem; consideration of the problem situation from several different viewpoints). In his fundamental survey [1] of different approaches to solving various problems, M. C. Jackson described in detail 10 such technologies used in situations that differ in a level of disagreement between stakeholders, and in a degree of exactness of the available model of real situation. Each technology is a specific (adjusted to the situation) combination of all the obligatory (for success) operations, described in the following text.

5.2.1 Stage One. Fixation of the Problem and Problem Situation

Abstract

The task of this stage is to formulate the problem and document it. The problem statement is generated by the client; the role the facilitator is to find out what the client is complaining about and what he is unhappy with. This is the problem of the client as he sees it.

The task of this stage is to formulate the problem and document it.

Thus, it is necessary to try not to influence his opinion and not to distort it. In conversation it is better not to submit a remark like "I agree (disagree) with you", but only "I'm listening to you". (This rule should be followed when interviewing at subsequent stages: after all, we need information not from the analyst but from those with whom he talks.)

The most serious mistake at this stage would be to immediately address the problem. This cannot be done for a number of reasons.

1. The client asked for help because he could not solve the problem on his own as the situation was difficult for him. Therefore, his model of the situation is inadequate, and he can only report the situation that he knows about it. Therefore, an attempt to solve the problem at once only with this information is doomed to fail. In fact, the next steps are designed to collect missing information to ensure the adequacy of the model.
2. Subsequently, it will become clear, including to the client himself, that the initial definition of his problem is inaccurate, or incomplete, or even incorrect. Often times the client diagnoses the problem and at the same time is mistaken. Symptoms are often mistaken for causes.

For example, when a patient comes to the doctor with a complaint of pain in the left hand, the doctor will fix the anamnesis, but will not immediately treat the hand, directing the patient to a cardiogram: perhaps this is an indication of heart disease.

Consider another example: the firm complained that the production departments do not have time to produce parts to customer order, and it turned out that the delay is excessively long in prior clerical and order processing.

3. In the course of the study, it may turn out that to eliminate the problem of the client, it is not necessary to solve it, but someone else's, a completely different problem.

   For example, employees of a large organization complained about the long wait for elevators in their multistoried building. Technical services offered the following options: (1) to build additional elevators, (2) to install high-speed elevators instead of the old slow ones, and (3) to create a single dispatching control of elevators. The psychologist offered to hang large mirrors in the pre-lift halls. The option turned out to be the cheapest and most effective: women engaged in preening, men quietly spied on them, and the complaints stopped.

   In another example, to solve a number of student problems, it is necessary to solve the problems of teachers first; to solve the problems of patients, it is necessary to solve the problems of medical staff.

Thus, fixing the client's problem is only a starting point, the beginning of a systematic study, and not a ready formulation of the problem to be solved immediately.

As for the fixation of the problem, it should be noted that the documentary design of the work done not only at this but also at subsequent stages is necessary in connection with the shortcomings of human memory, increase in the volume of information in the course of work, changes in the environment over time, etc.

Questions and Tasks

1. Why is it necessary to document the client's problem?
2. Why not start solving the problem immediately after fixing it?

5.2.2 Stage Two. Diagnosing Problems

Abstract

The task of this stage is to diagnose and determine what type the problem belongs to. Sometimes the solution to a problem lies on the surface. But often the diagnosis of the problem is not easy. An error in the diagnosis will lead to incorrect actions and will only cause harm. Although with our efforts to implement an improving intervention, the damage will be reduced.

Which problem-solving approaches to apply to solve a problem depends on whether we choose the impact on the most dissatisfied subject or intervention in reality, with which he is dissatisfied (there are cases where it is advisable to combine both effects). The goal of this stage is to diagnose and determine what type the problem belongs to.

Sometimes the solution to this problem lies on the surface (as in the case of a violently insane person who needs to be treated; or with an accident that needs to be eliminated; or with a conflict personality in the team). However, often the diagnosis of the problem is not easy. An error in the diagnosis will lead to incorrect actions and will only cause harm. Although with our efforts to implement an improving intervention, the damage will be reduced.

Diagnosis is a complex matter. Since it is difficult to give some general theoretical recommendations for implementing this stage, diagnosis is more art than science, intuition, experience, and luck play an important role in it. Yet, there are clues to how to do it.

For example, the English philosopher J. Mill advises: "Look for something that is common to every failure and that never appears in the case of success". In some cases, this advice may help.

Questions and Tasks

1. Try to formulate considerations that would help you choose between whether to influence the subject or to intervene in the problem situation itself.

5.2.3 Stage Three. Making a List of Stakeholders

Abstract

Our ultimate goal is to have an improving intervention. To subsequently take into account the interests of all participants in the problem situation (namely, the concept of improving intervention is based on this), it is necessary first to find out who is involved in the problem situation and to make a list of them. Due to the universal interconnection of everything in the world, we limit ourselves with only essential and most significant participants of the problem situation, called further stakeholders.

Our ultimate goal is to have an improving intervention. Although each stage should bring us one step closer to it, it is necessary to take special care that this step was exactly in the right direction and not in the opposite direction.

To subsequently take into account the interests of all participants in the problem situation (namely, the concept of improving intervention is based on this), it is necessary to first find out who is involved in the problem situation and make a list of them. It is important not to miss anyone: it is impossible to take into account the interests of someone who is unknown to us, and to disregard the opinion of someone threatens that our intervention will not improve. Thus, the list of participants in the problem situation should be *complete*.

Unfortunately, the task is impossible. Because of the openness of all systems, everything in the world is interconnected (a consequence of the second property of the system), and, therefore, the entire universe participates in the problem situation in one way or another (Figure 5.3), and rewrites an infinite number of its parts, which is an unthinkable thing. The way out is to describe the infinite variety of the universe in a simplified way — of course, through a *classification model*.

Indeed, although everything in the world is connected with our problem situation, it is connected to a different extent: some are close to it, others are far away;

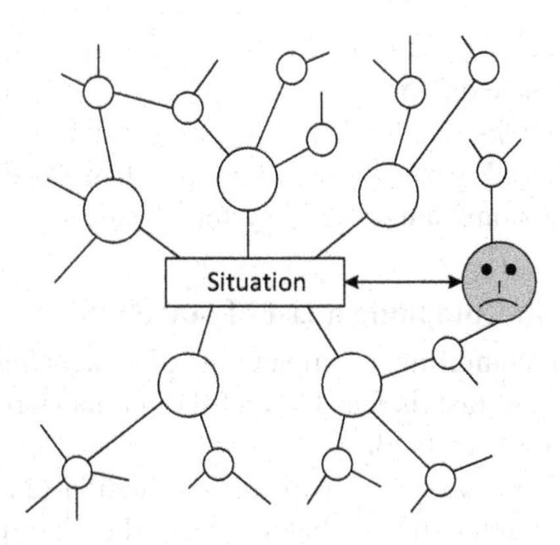

FIGURE 5.3 The whole world participates in any situation (but in various grades).

some are closely, strongly, and directly connected with it, while others are weakly or indirectly connected. Direct participants because of their indirect connections have information about the latter and somehow represent them in the situation. This allows the whole universe to be divided into two classes: the first class will only include the direct participants in the situation, while the second part will include all the others (Figure 5.4), confining ourselves to take into account only those who got into the first class. Therefore, in a certain limited neighborhood of the problem situation, there are now a finite number of elements, and their census becomes a real thing. However, the requirement of completeness of the list ("universal census") remains and even becomes tougher.

According to the classification canons described in Chapter 3 (Part I), each class must be given a name that all its members will carry. This can be done in any language. In English, it was done by analogy of Figure 5.4 with the situation at the racetrack: on a rectangular field, a horse race is going on, and coming at them the

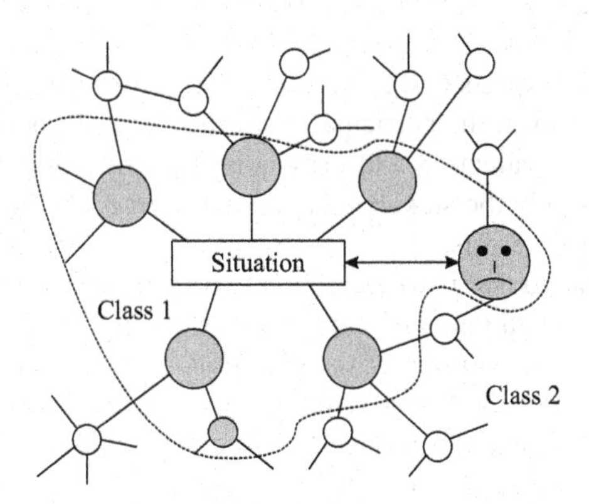

FIGURE 5.4 Stakeholders are essential (most influential) participants of the situation.

spectators place their bets on their favorite horses. The situation is one, the interests are different — a complete analogy with any other problematic situation. Players to tote are referred to as *stakeholders*. It was suggested that the same term should be used to refer to all "direct" participants in any problem situation. The term caught on, adding to the professional language of systems analysts.

5.2.3.1 Difficulties in Compiling a List of Stakeholders

So, we have the task of compiling a complete list of stakeholders of our problematic situation. In principle, the task is feasible and the number of stakeholders is finite. But, in practice, this task is difficult.

The main difficulty is associated with the evaluation (and, consequently, with subjectivity) of the characteristics of belonging to the class of stakeholders. Who else is considered "close", "directly related", and who is no longer? The boundary between direct and indirect participation should be drawn, but it is relative.

For example, the family of the loser in the sweepstakes a large sum—direct or indirect participant? Or a more important example, a list of stakeholders for an investigator of a crime. It turns out that except criminals and direct witnesses to the crime, to improve the reliability of the information in this list it is necessary to enable indirect witnesses: the latter did not know about the crime, but knew a lot about its participants and direct witnesses.

Similar problems arise in the economic analysis of the company's position in the market (which of the vast environment to make stakeholders?); or when designing a technical device (who will somehow deal with it — in its production, operation, trade, service, disposal, etc.?).

The guiding principle of stakeholder selection is that this class is the field from which information about the entire situation will be gathered. This information is needed to build an adequate model. So, the stakeholders include all those who have the necessary information.

However, the actual involvement of the subject in the problem situation is not a guarantee of the necessary information. Participants in a situation may identify problems associated with it, but designing appropriate sections of the improving intervention may require more in-depth, specialized knowledge of the specific issue. In such cases, you will have to resort to the services of experts and specialists in such matters. What kind of experts will be required will be revealed only at the fifth stage described below. However, the participation of incompetent, but direct stakeholders will be necessary for evaluating and criticizing the descriptions and projects proposed by experts, as only the stakeholder can determine whether these descriptions and projects meet his interests.

A very reasonable advice is to include an external observer in the team for the evaluation of the organization, who can perfectly listen to others and observe himself, along with an *internal* observer with tangible experience in the system. At the same time, it is recommended to take internal representatives from among employees with two or three years of experience in the company, so that they already see its shortcomings, but are not used to them.

5.2.3.2 Tips to Facilitate the Work

The experience gained during this stage can be arranged in the form of hints, heuristics, useful tips, which can increase the completeness of the list of stakeholders. Here are some of these tips found in different sources.

1. *List of stakeholders is a black box model for the problem situation.* We are waiting for the errors of the first, second, and third types, discussed in Chapter 2 (Part I) and we need to take any available measures to prevent them. But this is not an easy task.

   For example, in the 1970s, when problems were discovered in USSR's national economy, the output was seen in the improvement of management, in particular, in the program-target management. To this end, all agencies built "target trees", on the basis of which short-, medium-, and long-term plans were developed. For example, a group of economists from Moscow did this work for the Ministry of the Navy of the USSR. They took the black box model of the Ministry as a basis. The scheme is simple: it is necessary to consider below, above, and standing next to the system. The higher systems were obvious: the Central Committee of the CPSU and the Council of Ministers; it was decided to take the coastal fleet and the long-distance fleet as the lower ones; the fleets of socialist and capitalist states were called standing systems. According to the existing methods a target development program of the merchant marine was compiled. In 1984, the media began to report the collapse of the system. The result did not match the expectations — a sure sign of the inadequacy of the model. On closer examination, errors of the second kind are detected: for example, railway and river transport were not included in the number of stakeholders. But the main mistake was not taking into account the interests of the system itself, as a result of which there was an acute shortage of housing, children's, medical, and cultural institutions in the infrastructure, which led to the withdrawal of qualified personnel, even ship captains. These are the consequences of ideology, which believed that their own, and even more personal interests are insignificant, something like the "remnants of capitalism".

   Hence, "black box" advises *whose* needs to be done, and *how* it will be done depending on the analyst (facilitator).
2. *"Silent stakeholders."* Often, the number of stakeholders should include not only subjects (individuals, groups, organizations) but also other participants in the situation. At the end of the 1980s, at the International Institute of Applied Systems Analysis (Vienna, Austria) under the leadership of R. Akoff a "round table" on the theme "Art and science of system practice" was held. One of the many results of this conference was a remarkable recommendation to increase the completeness of the list of stakeholders through the mandatory inclusion of three "silent" stakeholders:
   1. future generations (they do not exist yet, but their interests must be taken into account, so as not to create problems for them by our intervention

in today's reality, as did previous generations to us — debts, exhaustion of renewable resources, the problem of nuclear and industrial waste, acid rain, etc.);

2. past generations (they no longer exist, but their interests are represented by the culture they left to us. Intervention cannot be recognized as improving if it causes at least some damage to material or spiritual culture);
3. environment (intervention cannot be considered as improving if it harms our environment, both living and inanimate).

How exactly the interests of silent stakeholders will be taken into account in your intervention depends on the nature of the problem and how deeply the developers are imbued with the ideology of improving intervention.

3. A list of *obligatory participants.*
   a. In every situation, there are those who get something in it, buy and use something. Our intervention can change their position and opportunities, and we promise to implement an improved intervention that does not infringe upon their interests.
   b. In each situation, there are those who work by performing some actions. Our intervention should not harm them. We need to add the actors to the list of stakeholders.
   c. In any situation, some organizations, enterprises, and institutions are involved. Changes will be made according to the interests of some of them ("problem-containing" ones), carried out at the expense of the resources of others ("problem-solving" ones), and somehow affect the others. There is no doubt that if a leader from any organization does not like our intervention, he will use his influence, connections, and resources to prevent this. Our task is to make each of them an ally or sympathizer, at the very least a neutral observer, but no one an opponent. Otherwise, the intervention will not be improving.
   d. Any situation involves material resources: land, water, buildings, structures, mineral reserves, etc. And they all belong to someone: the state, groups of people, and individuals. Intervention in a situation will inevitably affect their interests, and we intend to harm no one. Therefore, the owners must be listed.
4. *Tip of the European Commission.* Another hint was proposed by the European Commission in line with the recommendations for those planning to receive a TEMPUS/TACIS grant for a project. Here is the relevant section of it:

   "The following questions can help you determine who are stakeholders:

   - What do you (who make up the plan) need to know? Whose opinion and experience would be helpful?
   - Who will make decisions on the project?
   - Who is supposed to be the executor of these decisions?
   - Whose active support is essential for the success of the project?
   - Who has the right to be a participant in the project?
   - Who can see a project as a threat?"

To answer these questions in relation to your situation, you need to include the following as stakeholders: (1) the participants of the situation you need as experts, (2) representatives of problem-solving systems, (3) representatives of problem-containing systems, (4) those desirable to have an assistant or ally in the project, (5) subjects legally related to the situation, and (6) those whose careless (nonimproving) intervention may have an adverse impact.

Though using any or all of the above tips will increase the completeness of your list of stakeholders, it will not be exhaustive. Perhaps, at later stages, it will be revealed that someone significant is missing from the list, and you may have to go back to this stage and add them to the list.

Questions and Tasks

1. Who are the stakeholders?
2. Does it mean that in the future we will only account for the interests of stakeholders and neglect interests of all others?
3. Do you remember hints helping you to make more complete list of stakeholders?

5.2.4 Stage Four. Revelation of the Problem Mess

Abstract

Purpose of this stage is revealing assessment of problematic (for the client) situation by each stakeholder.

Stakeholders have interests that we need to take into account. But for this they need to be known. In the meantime, we have only a list of stakeholders. The first piece of information that needs now to be obtained about the stakeholder is his own assessment of the situation, which is problematic for our client. It may be different: some stakeholders may have their own problems (negative assessment), some may be completely satisfied (positive assessment), while others may be neutral to reality. So the "expression of the face" of each stakeholder must be clarified (Figure 5.5). In fact, we have to do the work that we did in the first stage with the client, but now with each stakeholder individually.

R. Ackoff [2] suggested calling the resulting list of the subjective assessments of the existing reality (which is a problem for the client) the *problem mess*. Although this term has a slight slang shade (which is why some authors prefer to call it more dryly "*problematique*"), it successfully emphasizes a very significant and fundamentally important point: its judgments are not independent, but are intertwined, interrelated (as being judgments about the same). This means that they form a complete system, with all the ensuing consequences.

In this case, the main consequence is the property's indivisibility into parts (the tenth property of the system). Hence, it becomes obvious that not inadmissibility (unfortunately, it is not always possible to keep someone from doing stupid things), but the undesirability and incorrectness of solving one, albeit a very important problem in isolation from other components of the mess.

Thus, the problem of the client in the mess is that it reveals as its core, the germ, around which the views of other stakeholders are grouped. Now it is clear that our task

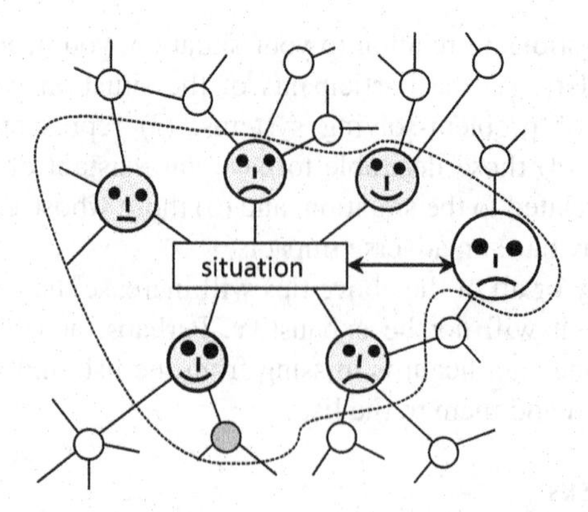

FIGURE 5.5 The mess is a description of the situation assessments of each stakeholder.

is not to solve the client's problem but to work with the problematic mess as a whole. This is where the design and implementation of improving intervention, which solves the client's problem, taking into account the interests of all stakeholders, serve.

5.2.4.1 Technologies for Identifying the Mess

J. Warfield [3] developed a specific technology for developing a problem mess that has been successfully used in practice.

The first stage of his methodology, the group wording method (NTG — nominal group technique), reveals the problematic mess by posing to the group of stakeholders the question: "What problems do you see in this problem situation"? Each of them creates a written statement of the problems they recognize, and their printouts are posted on the wall for general viewing. After 15–30 minutes, the generation of new ideas fades. Then a discussion of each of them is held to clarify what exactly the author has in mind. Such a discussion may take two to three hours since the number of problems usually ranges from 40 to 160, and the discussion may become a serious debate at some topics.

However, the practice of solving the real-life problems of large organizations shows that identifying their problematic messes often cannot be reduced to a one-time, single interview of stakeholders. Each participant in a problem situation looks at it from its local position and sees only one of its sides, and hence, only the details of this side are contained in his model of reality. This is not enough to adequately describe the situation. In such cases, the identification of the problem mess becomes a thorough study of the situation, the building of a detailed, rich picture containing the fullest possible information about the actual problems associated with this situation.

The limited reality description by the subject is due to the fact that he perceives reality only through his models. Therefore, an increase in the flow of necessary information can be achieved by offering the subject to look at the situation not from one (usual for him) viewpoint, but by using other models.

The technology of systems analysis uses different ways to implement this idea. For example, the division of stakeholders into two groups was applied. Supporters of the problem's solution "from top to bottom" and "from bottom to top", that is, those who prefer to plan from a global goal and decomposing it to the lowest-level goals, and those who are trying to aggregate the lowest-level goals to synthesize a global goal. Formulating the supposed problems of each stakeholder is proposed to these groups, followed by a dialectical discussion to work out the final wording of the problematic mess.

Another variant of the proposal for a set of different ideas about organization (metaphors) is as follows: organization can be viewed as a "machine", "organism", "brain", "culture", "political system", "suppression tool", and so on. Each of these viewpoints highlights some features that are not visible from other positions, as well as some aspects that are common. This allows us to see the limitations of the model, which we adhered to at the beginning, and to expand in more detail the problematic mess. The publications report on practical utility and other metaphors, for example, "organization as a continuously changing system", "like a madhouse", and "like a carnival".

The "machine" metaphor draws attention to the purpose (goal) of the system, to the composition of the parts, their functions, and the connections between them, usually expressed by hierarchical structures. The "organism" metaphor describes a system whose primary goal is survival in a turbulent environment.

The metaphor "brain" highlights the importance of information processes, decision-making, governance, training, and correction of goals. The metaphor "culture" concentrates on the individual characteristics of workers, their values, and personal aspirations, as well as on a corporate culture that unites the team. The metaphor "political" focuses on relationships between people in organization, relations of power and responsibility, rivalry and cooperation, conflict resolution, etc. Metaphors of the "tool of repression" and "madhouse" focus on the negative aspects of life in an organization: restriction of freedom in thoughts and in the disclosure of capabilities, exploitation, prevalence of punishments over rewards, and discrimination on some grounds. Each metaphor is compared with a system model, and the problem is identified by describing the system in question in terms of this model.

A community of systemic organization of everything in nature allows to look for analogies not only between metaphor and our system but also between our system and *any* entity in the world. "Method of random associations" suggests to look for analogies between our system and several others arbitrarily chosen from dictionary nouns.

Another variant of disclosing the diversity of approaches to the problem situation in management is to indicate the presence of different paradigms. The word "paradigm" denotes a vision of the world, a set of ideas, assumptions, and beliefs that guide people in their actions.

The difference between metaphors and paradigms should be emphasized. A metaphor is a partial, one-sided view of an object or phenomenon. Different metaphors do not contradict but complement each other. Proponents of different paradigms believe that they offer the best way to describe the observed "reality". Therefore, paradigms are incompatible, and so are the descriptions they generate. Therefore, the recommendations given to the manager by experts who adhere to different paradigms are contradictory, which forces him to look for something "average".

There are four paradigms: functional (the functionalist paradigm), explanatory (the interpretive paradigm), liberal (the emancipatory paradigm), and postmodern (the postmodern paradigm).

The "functional" paradigm proceeds from the fact that with the help of scientific methods one can understand how the system works by ascertaining the nature of the system parts, the interaction between them, and between the system and its environment. The knowledge gained will help the manager to improve the organization's management. Metaphors "machine", "organism", "brain", and "changing system" are usually associated with this paradigm.

The "explanatory" paradigm believes that organizations are social systems created to achieve subjective goals resulting from the interpretation of the circumstances in which the subject finds itself. An organization is created by people, and people act in it in accordance with their interpretations of reality. This paradigm aims at achieving an understanding of the meanings that organizations bring to the joint activity, determining the overlapping areas of these meanings, and thereby creating a common purposeful work. Thus, managers are focused on achieving the necessary level of the overall corporate culture in the organization, making decisions with employee participation, and increasing their commitment to the organization. Cultural and political metaphors are usually associated with this paradigm.

The "liberation" paradigm aims to "liberate" individuals and groups in organizations and society. It is wary of the authorities and tries to publicize its methods and concrete facts of suppression and coercion, which it considers illegal. It criticizes the status quo and calls for radical reforms and even a revolutionary change in the existing order. It is against any form of discrimination (by race, gender, status, age, etc.). This paradigm is often associated with the "tool of repression" and "madhouse" metaphors.

The "postmodern" paradigm is in opposition to the rationalism of all three modernist paradigms. It believes that social systems are so complex that attempts by other paradigms to give them a rational explanation are useless. In particular, they cannot explain aspects of pleasure, entertainment, and emotions in the actions of people in an organization. It insists on open conflict resolution, freedom of expression of opposing opinions, and encouraging diversity and changes. This paradigm corresponds to the carnival metaphor.

So far, we have discussed about the "problem mess" as a kind of "photographs" of the current state of relations of the stakeholders to the existing situation. However, everything that happens with time is changing, with a lot depending on what happens in the future. Therefore, for a truly "systemic" solution of a problem, it is necessary to rely not only on information about the current state ("photo") — a static model of the system — but also on its dynamic model ("movie").

Thus, the formulation of the problem mess requires identifying a set of interrelated threats and opportunities for an organization or institution, the problem that we have undertaken to solve. It is this aggregate that is the complete picture ("rich picture") of the problem situation, the "mess". It determines how the organization would crash if it continues to act in the same way as so far, that is, if it cannot adapt to changes in the internal and external environment, even if it could accurately predict the course of these changes. This reveals what the organization or institution should avoid at all costs.

Such an approach to formulating a mess is almost the same for an organization that is already in crisis, and for one that is only alarmed by negative trends and wants to prevent a looming crisis.

Sometimes the formulation of a mess can be reduced to identifying the problems of the stakeholders, which can be done by brainstorming at a session of the stakeholders or their representatives. But with a sufficiently large complexity of the situation, it may require a more detailed consideration of it. For example, R. Ackoff [4] recommends doing it in several stages.

1. Perform systems analysis. This is a detailed description of how the organization or institution is currently operating. It is convenient to present it with a series of flowcharts showing how input material is acquired and transformed in an organization, and how money and information flow in it. These flowcharts can be prepared separately, but it is usually useful to combine them into a single scheme or to depict them on transparent slides; when overlaid on them, their interconnections are easily viewed.
2. Perform obstacles analysis. Identify those characteristics and properties of the organization that hinder its progress or hinder changes (e.g., conflicts or traditions).
3. Identification of scenarios (reference projects) of a possible future. Formulate what the course of events will lead to, and what is the future of an organization if there are no changes in its existing plans, programs, policies, and practices. This should expose the possibility of self-destruction of the organization, and show how the obstacles described in stage 2 will prevent the necessary changes to be made.

5.2.4.2 Structuring the Mess

Individual views of the stakeholders on the problem situation, recorded in the minutes of interviews with them, are the initial data containing information about the system under consideration, still in an implicit form.

Extracting information is expressed in the construction of more and more meaningful models (*data, information, knowledge, understanding, wisdom* — see Chapter 3, Part I).

The first step is to sort, group, and classify data (results of observations and measurements). This initial processing gives a description, called by R. Ackoff *information.* Considering the stage of the formation of a mess, this operation leads to the formation of groups of problems in the static description of the mess and to the emergence of scenarios (reference projections) in the dynamic one.

There are technologies that do not require further structuring of the problematic mess (e.g., idealized design). However, there may be a situation when it is necessary to proceed to considering the interrelations between the components of the problematic mess, and to build models of the next level, such as "*knowledge*".

A typical example of this is the situation when it is clear at the start that the available resources are not enough to solve all the problems that make up the mess.

There are questions about which problems to give preference and how to take into account all the others correctly. The answer is to design a sequence of improving interventions for different resource allocations for mess problems, and from all such options, to choose the one that gives the greatest improvement under the given constraints (i.e., optimal). Usually, among the limitations, in addition to the volume of resources, there is a mandatory improvement in the situation for the client, but at the same time, it may turn out that the main resources do not have to be directed to personal problems.

The task of allocating resources and determining the priority of solving problems is greatly facilitated if we manage to somehow sort the mess. However, this is not always possible. An example, when possible, is planning a targeted project, in which each participant must identify problems for themselves and fulfill their role. In this case, after identifying the mess, it is recommended to start building a hierarchical problem tree, thereby revealing the cause–effect relationships between them.

One variant of the problem tree construction technique can be described as follows. Stakeholders (project participants), having familiarized themselves with the mess, write out, at their own discretion, a key, focal problem in the entire mess. Everyone is guided by their own interests and their problems. Then there is a collective discussion of the ranking of problems until the participants agree on which problem is the starting one. Subsequently, the next problem is taken and compared with the first one. Then:

- if the second problem is the cause, the condition for the first, it is placed a level below;
- if it is a consequence of the first, it is placed higher;
- if it is neither a consequence nor a cause, then it is placed on the same level.

This scheme addresses the remaining problems. In the course of work, it may be appropriate to change the focal problem, but this does not deprive the meaning and significance of the analysis.

For example, if the focal problem is "Insufficient number of qualified professionals in such and such a specialty", then its reason can be formulated as "Insufficient level and quality of higher education in this specialty".

In the presence of a problem tree, the question of the sequence of problems is solved in an obvious way: the higher problems cannot be solved until the lower ones are solved.

However, cases where problems can be built into an ideal tree-like hierarchical structure are rare. Because of the general interconnection in nature (a consequence of the second property of the system), more often, only the essential connections between the components of the mess form a network-like structure, in which the mutual influences sometimes form closed feedback loops. This is what leads to the understanding that for complex problems there are no simple solutions, and that any problem has no single reason, although among them there are "more important" and "less important" ones.

In general, even when the mess is not tree-like perfectly, the sequence and intensity of problem-solving should be determined precisely by the structure of the mess, the desire to move the whole mess to improvement, and not to those of the stakeholders who are more persistent in seeking to solve their own problems.

J. Warfield [3] proposes a different approach to structuring a problem mess, that is, to build not a tree-like but a network structure. He calls this "explanatory model building" (ISM, interpretive structural modeling).

For each pair of problems, the question is: does the problem "x" exacerbate the problem "y"? The answer is the basis for including the edge between "x" and "y" in the graph and applying a direction to the edge. The wording "x" and "y" may need to be clarified. The decision on the existence of a relationship between "x" and "y" and its focus was adopted by a vote of the majority ("don't know" equates to "no"). Consequently, we obtain a directed graph describing the structure of the mess.

For a large number of pairs in question, this work can be done with the help of computers (the corresponding programs can be downloaded free of charge from http://www.gmu.edu/depts/t-iasis/).

The resulting graph can be further transformed into a smaller number of simpler second- and third-order coarsened structures, grouping the problems by some features. For example, the author recommends to group the problems according to their importance. Each stakeholder is invited to choose from five major problems and continue to work with them, for example, analyzing their overlap.

The ultimate goal of structuring is to transform the graph so that the arrows are directed from left to right, highlighting the problems that should be solved first and those that can be solved later. Of course, there are difficulties in the case of cycles in the graph, which are subject to separate consideration.

Structuring the mess is a preparatory step to a consistent solution of a set of all problems.

However, the method of successive, small, and "continuous improvement" is criticized by many as a half-measure, poorly coordinated with the variability of the environment, without fully taking into account the emergence of the system.

When we remove the unwanted, we often obtain even worse one. Getting rid of a television program by switching to another channel, we get to a program that we like even less. The use of dust against harmful insects was harmful to the environment. The legal prohibition of alcohol gave rise to organized crime. Prisoners in jail often only increase their criminal skills. Therefore, getting rid of the undesirable is often just a departure from responsibility.

As an alternative to the incremental approach, the method of "idealized design" by R. Ackoff deserves special mention. This technique does not require the ordering of the problem mess: the identification of problems and their consequences are only needed to understand the need for change. The subsequent design of changes comes from the assumption that the existing (and not satisfying us) system "disappeared last night", and the work is not aimed at correcting undesirable features of the organization, but at the design of the desired.

5.2.4.3 Participation of Stakeholders in the Analysis

The best source of reliable, accurate, and complete information about the stakeholder is, of course, he himself. To find out what his problems are, you need to contact him and talk to him. So we faced the need to involve stakeholders in the process of analysis. (It will later become clear that this is necessary for the entire process, not just at this stage.)

No one can express their opinion better than they do. But getting first-hand information faces the *problem of stakeholder availability*. "Silent" stakeholders are by nature noncontact; some participants in the situation do not have the opportunity or desire to cooperate with the analyst; some stakeholders may not be a person, but a group, and even numerous; and some may not be personally available for geographical or political reasons. But information about them is still needed!

The problem of inaccessibility can be solved as follows. All stakeholders are divided into two types: *obligatory* and *desirable* participants of the analysis.

First, you must determine problem-keeping and problem-solving systems among organization's stakeholders. In the problem-holder system, the question is clear: it is the representative who asked the analyst to become his client. Next, it is necessary to identify the problem-solving system(s), that is, those at whose expense and with whose help the intervention will be implemented. Often the same organization is both problem-holding and problem-solving (e.g., a powerful firm), but sometimes they differ (e.g., school and the city council, municipal and federal authorities).

It is of fundamental importance that the first persons of the problem-holding and problem-solving systems are obligatory participants of the system analysis. Without their participation, all the work has no practical meaning, which makes it hopeless: after all, they have the resources and powers necessary to solve the problem. That is why the guarantee of their participation is included in the contract, and if this condition is not accepted, the analyst does not take up work.

Other stakeholders are desirable participants. That is, it is best if they can be included in the collective process of systems analysis. But if someone is inaccessible, it is possible to obtain the desired information about him and without his personal participation. This possibility is based on the consequence of the second property of the system — the universal relationship in nature. There is a chain of subjects between the analyst and the inaccessible stakeholder. Being related, they contain information about each other, the greater the closer they are to each other. Therefore, we can try to obtain information about the inaccessible stakeholder to get in touch with those nearest to him in the chain and "extract" the information we need from them instead. This, of course, is less reliable than the stakeholder, the source of information about it, but we have no better opportunity. The last thing is for the analyst to express the opinion of the stakeholder: he is the most distant link in this chain. This is only allowed if everyone else in the chain is unavailable.

The task of finding a sufficiently knowledgeable representative of an inaccessible stakeholder is not as difficult as it seems. Probably everyone had to talk to a stranger to find that you have common friends. There was even a scientific study of this issue by scientists at the University of California. The target person was represented by a student of this university. Then, from the list of voters of the most remote from California, Pennsylvania, was randomly selected 100 people who agreed to participate in the experiment. Each of them was given a letter addressed to the girl, but it was possible to send it directly to her only on condition that the sender knows her personally. If not, the letter was sent to one of his personal friends on the same condition. The U.S. mail provides a service of correspondence through intermediaries, so that the letters eventually reached the addressee. The length of a particular chain was determined by the number of post stamps. It turned out that in order for two strangers out of 200 million people contacted by a chain of people who know each other, you need quite a few links. The shortest chain was two (a farmer has his childhood friend a Congressman in Washington, and the Congressman's friend is a professor at that university), the longest was 11, and the average length was 4.5. It should not be surprising that with an average of 100 acquaintances, a network of chains of four people will cover 100 million individuals.

So, in case of unavailability of the stakeholders, it is necessary to find their best possible representatives and involve them in the team working on the solution of the problem. In particular, representatives of "silent" stakeholders can be experts, scientists, heads of relevant administrative bodies, and public organizations of relevant profiles.

It remains to say that if the stakeholder is a group of people (students of the city, pensioners, entrepreneurs, etc.), then either try to attract a competent representative of the group (among its leaders), or resort to the methods of applied sociology and statistics (to clarify public opinion).

Hence, we assume that the selected, instructed, and inspired by the systems analyst group of stakeholders (or their competent representatives) formulated the mess and structured it as possible.

Questions and Tasks

1. What do you call a "problem mess"?
2. What is the dynamic version of the problem mess?
3. Why not solve the problem of the client in isolation from the problem mess?
4. What does it mean to "work with a mess as a whole"?
5. How to solve the difficulties arising from inaccessibility of the stakeholders?

5.2.5 Stage Five. Definition of Configurator

Abstract

*A **configurator** is a minimal set of professional languages that allows you to give a complete (adequate) description of a problem situation and its transformations. We emphasize that the configurator can help decide which third-party specialists should be involved in solving our problem: those whose professional languages from the configurator are the stakeholders themselves and the analyst himself speaks insufficiently professionally for our purposes. Such a situation would require external experts.*

A prerequisite for successful solution of the problem is the presence of an adequate model of the problem situation: with its help it will be possible to test and compare the options for the proposed actions. This model (or set of models) must inevitably be built using certain language (or languages). The question is how many and which languages are needed to work on this problem and how to choose them.

For example, in case of a traffic accident, then the resolution of the problem may require different languages: legal (who is responsible for what), medical (condition of the road accident participants before and after), technical (condition of the road and equipment), administrative (organization of all consequences), economic (financial security), etc.

It is important to emphasize that real-life problems are not single-disciplinary, that is, described in the language of only one specialty. Only learning tasks can be single-disciplinary, and even then not always (e.g., physical tasks require knowledge not only of physics but also of mathematics and other colloquial languages).

A *configurator* is a minimal set of professional languages that allows you to give a complete (adequate) description of a problem situation and its transformations.

All work in the course of solving the problem will occur in the languages of the configurator, and only on them. Defining configurator is the task of this stage. We emphasize that the configurator is not an artificial invention by systems analysts but invented to facilitate their work.

On the one hand, the configurator is determined by the nature of the problem. Take, for example, the geometric case. Suppose there is a line on which a point is marked (Figure 5.6a) and it is required to describe where it is. For this, we need a language in which we describe it. The elements of the language are a certain point of origin 0 and a unit interval (Figure 5.6b). Having introduced grammar and syntax (operations of postponing unit intervals and working with them), we can say that the point of interest lies in the fifth unit interval (Figure 5.6c).

If necessary, more accurate statements are introduced by the fraction of a unit interval. Thus, our configurator for this problem consists of one language. If you need to describe the position of a point on a plane, you will have to build a configurator from two languages, and for a three-dimensional task from three languages.

On the other hand, the configurator can be viewed as another property of the system by which the system solves its problem. For example, two eyes and two ears are given to us as the material carriers of the configurator for determining the location of the source of light or sound on the plane. The third language, for solving volume problems, is the possibility of turning the head and, accordingly, changing the orientation of the plane of definition. Dragonfly, unlike a bat, cannot turn its head when chasing midge, so nature has put on its "forehead" a triangle with small eyes in its corners.

Let's return to our main task of defining the configurator of our problem situation. Practitioners are often guided by intuition, commonsense, experience, and expert advice. Like any subjective decision, it may turn out to be true, but it may also contain errors. At the same time, we already have an objective information that does not depend on someone's opinions for building the configurator of our problem.

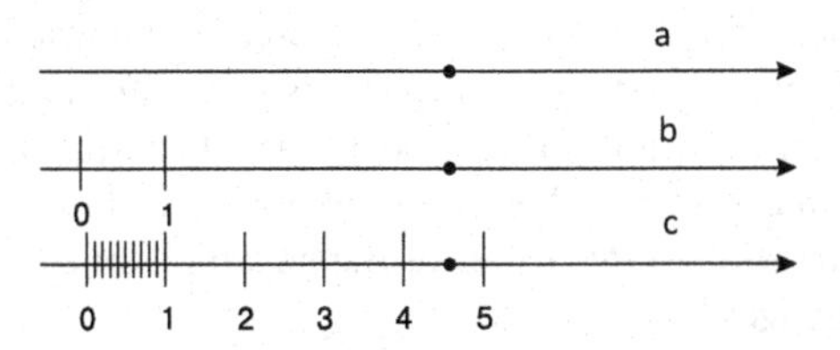

FIGURE 5.6 Languages of different exactness for describing the place of a point.

However, in an implicit form, it is contained in the minutes of conversations with stakeholders about the problem situation. The fact is that each of them spoke only about what he considers important, that is, spoke the languages of his configurator for this situation. Therefore, we have the opportunity to "calculate" his configurator by analyzing its text from the problematic mess, but not in terms of what he was talking about, but in what languages he spoke. Someone paid attention to financial aspects and discussed health problems. In its configurator, there are economic and medical languages. For other mentioned legal issues and relationships with other people, legal language and the language of psychology are part of its configurator; and so with every stakeholder. Consequently, we will have a set of configurators of all stakeholders. The configurator of the situation as a whole is their union. It may include several, and maybe many, languages. There can be a common language that everyone speaks, on the other, the majority, on the third, a minority, or even only one stakeholder.

We have to use them all; no language can be thrown away; otherwise, the relevant aspect will not be taken into account, which will not allow claiming an improving intervention. Of course, it would be unwise for all stakeholders to build models in all languages. When designing an improving intervention, it is necessary to tailor it individually for each stakeholder by building and using models only in the languages of its configurator.

A good example of this is the story of the inclusion of an astronaut from Malaysia in the crew of the International Space Station. The young astronaut, perfectly prepared for space flight, turned out to be a Muslim, which required the inclusion of the language of Islamic culture in his configurator. It instructs the devout Muslim to pray five times a day. The question arose how often and at what moments should he perform a prayer if the ISS orbits around the Earth 18 times a day? Further, the prayer is supposed to be performed by the person oriented by head toward Mecca. It demanded to create a special program so that the computer determines the direction to Mecca at the right time. Finally, the hierarchs of the Islamic faith had to allow an astronaut to use paperwork after the toilet, since under the conditions of weightlessness the Muslim custom of washing away with water is not feasible.

We emphasize that the configurator can help decide which third-party specialists should be involved in solving our problem: those whose professional languages from the configurator are the stakeholders themselves, and the analyst himself speaks insufficiently professionally for our purposes. In such cases, external experts will be required.

The need to attract information from various areas of knowledge creates a serious difficulty in the practice of systems analysis: it is necessary to use the languages of the configurator, which are not understandable to all participants in the development of improving interventions. The situation becomes similar to the biblical parable about the construction of the Tower of Babel, which failed because the builders spoke different languages.

The problem of communication of all members of the group can be solved in different ways. For example, the integrated science and technology group, which developed the automated control system for the Tomsk region's economy, included economists, cybernetics, engineers, lawyers, philosophers, programmers, managers, and administrators, and spent about a year of heated discussions about the essence of the project until the group members developed something like a common language and began to understand each other. Such a common language is based on sharing the different representations of information: verbal, graphic, mathematical, tabular, etc. An important role was played by a clear and explicit understanding which particular models are being considered at the given moment, static or dynamic, an analytical or synthetic consideration of the issue is carried out.

Questions and Tasks

1. Why is it necessary to define a configurator?
2. How to determine the configurator of a separate stakeholder?
3. How to work with configurators of different stakeholders in the design of improving intervention?

5.2.6 Stage Six. Revelation of Stakeholders' Purposes

Abstract

Total non-negative assessment of improving intervention by all stakeholders is possible, if it doesn't move them away from their purposes.

To strive for the implementation of improving intervention, we must ensure that none of the stakeholders consider the intervention to be negative. People give a positive assessment of change if it brings them closer to the goal, and negative if it moves them away from it. Therefore, to design interventions, it is necessary to know the goals of all stakeholders. Of course, the main source of information is the stakeholder himself.

We again come to the need to have an interview with each stakeholder. Though the work will be similar to what we did to figure out their attitude to the existing situation, we will now ask them about what they would like to achieve. As a result, by analogy with the problematic mess, we will determine what can be called a *target mess*. Knowing it will allow you to design an improving intervention. Let us explain this using the scheme in Figure 5.7.

Let the situation be characterized by two parameters — "q_1" and "q_2". The assessment of the situation by the stakeholder will be indicated on the diagram by a point. The problem mess is depicted by a group of points in the lower left corner of the scheme, and the target mess is in the upper right. It is now clear that any progress (positive change in the situation) along the path that brings us closer to the target

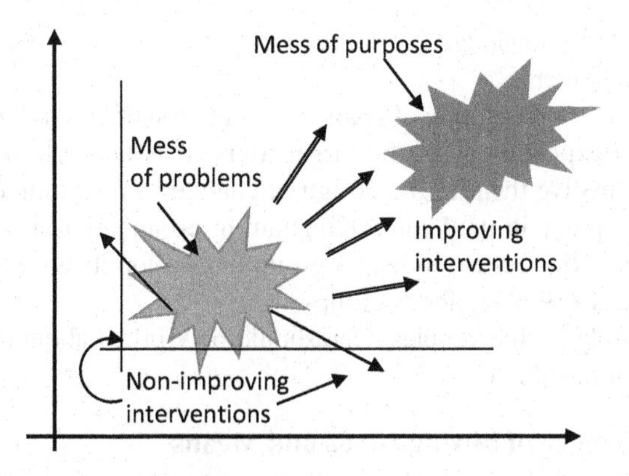

FIGURE 5.7 Improving intervention is any approach to purpose.

mess is an improvement. There are many such interventions. You can stop at any of them, but you can try to find the one that gives the greatest improvement.

In identifying the target mess, we face serious difficulty. If there are erroneous, untrue goals in the target mess, their subsequent implementation will naturally cause discontent and disappointment, and the resulting intervention will be not improving.

The difficulty lies in the fact that the goals declared by the stakeholder may be different from his true goals. It is unlikely that we will face deception or concealment of goals: the stakeholders understand that it is in their interest to help with the implementation of improving interventions and that they will harm themselves by giving incorrect information. But even with full and voluntary cooperation with and in good faith trying to properly state their wishes, the stakeholder may have difficulty and make mistakes. There are several reasons for this, and our task is to get to the true goals, taking into account all of them. We will discuss the main reasons for the discrepancy between the declared and true goals and some ways to overcome them.

5.2.6.1 Danger of Substitution of Targets

Sometimes substitution of goals of some stakeholders with the goals of others occurs.

Such a situation usually arises when professional specialists involved in solving problems impose their vision of the world and thereby replace stakeholder's main goals with their own. "The operation went brilliantly, but the patient died" is not a bad joke, but a statement that is common among surgeons. Consider another case when an elegant building was built on the campus of the University of Sussex (England), for which the architect was awarded the Gold Medal of the Royal Society of Architects in 1965, but its internal layout was unsuitable for educational or administrative purposes. Many award-winning commercials and posters did not have any impact on the company's business (recall the excellent advertising of the Imperial Bank in Russia of 1990s). A survey of the National Health Service in England found that less than 1% of the training time for doctors of this service was devoted to preventive medicine, although the organization was created for this very purpose. It should also be emphasized that the systems analyst, considering himself an experienced and knowledgeable professional, may begin to express his opinion instead of a stakeholder.

Another option of substitution of goals is the creation of "focus" groups of experts to determine the expectations of the target users. It is better to address the users themselves and involve them in the design of changes. For example, the leadership of a conservative party in the United Kingdom hired an environmental activist, an ardent opponent of the Tories, to lead a group to develop its environmental policy. The Labor Party is still using focus groups.

You can multiply such examples. One should be vigilant about this phenomenon during target-manifestation.

5.2.6.2 The Danger of Mixing Goals and Means

All the many goals that a subject proposed throughout his life include goals of different importance, long-term, scale, reachability, etc. The goals of the subject are not chaotic or independent. They are connected, coordinated, ordered, and forming a tree-like structure (Figure 5.8).

The peculiarity of this tree is that each of its elements has a dual meaning: it is a means for the upper-level element associated with it and a goal for the lower-level elements. Therefore, when a stakeholder is faced with the question of what his goal is in a given situation, he must decide which floor of the tree he is on. This is where the possibility of error is laid — choose a tool instead of a target, making a mistake on the floor.

For example, consider a case when city authorities appealed to the systems consulting firm to determine where it is best to build a new hospital. It was possible, after coordinating the criteria with the customer, to compare the selected sites with their help and to solve the optimization problem by satisfying the client. However, systems analysis recommends verifying goals for truth. The question: "Why do you need a new hospital"? is answered by "To improve the medical care of the population" (this was the true goal). Other means of achieving the goal were considered, and it turned out that for the same subsidies, it was much more effective to coordinate and modernize the network of already existing medical institutions instead of building a hospital.

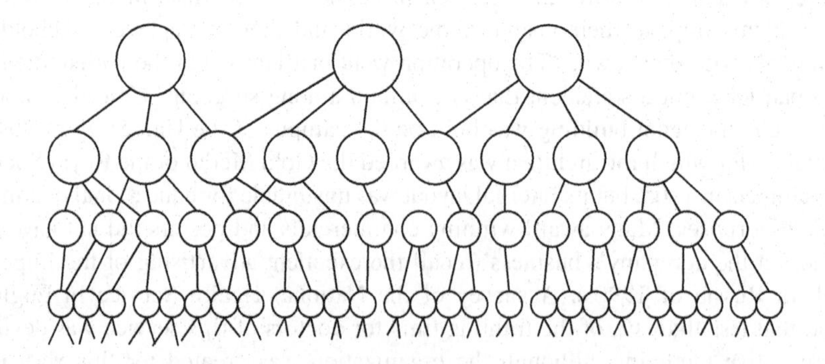

FIGURE 5.8 Goals of a subject are coordinated in a tree-like structures.

The technique which makes it possible to detect confusing the goals with means consists of asking the question: "Why do you need it"? If the answer mentions other goals, it is necessary to determine whether they belong to a higher or lower level compared to the declared one.

5.2.6.3 The Danger of Incomplete Enumeration of Goals

In many cases, the desired future is complex and its description consists of not one, but several goals. A very dangerous case is when the subject does not enumerate all of them due to either forgetfulness or finding some of them unimportant. Even if the goals mentioned are true, implementing an incomplete set of a multiple goal gives an absolutely unacceptable result.

For example, in one of the areas it was instructed to develop a proposal to improve the economic efficiency of the woodworking industry. In particular, it was proposed to merge small enterprises into one large association with obvious technological and economic benefits. Local authorities, however, blocked this project despite the fact that it fully met the specified criteria. It turned out that with the formation of a large association, the woodworking enterprises will be transferred from the system of local industry to the subordination of the union ministry. Although revenues will indeed increase, contributions to the local budget will decrease, as will the percentage of materials and products produced and distributed in the region. Thus, the incompleteness of the target display crossed out a lot of work to create the project.

In engineering folklore, there is a whole series of jokes under the general title "Correctly formulate a technical task".

These are fictional funny stories that illustrate how absurd the result is if you precisely follow an incomplete set of specified requirements.

Unfortunately, it is difficult to give recommendations on how to achieve the completeness of the composition of a multiple goal besides persistent repeated reference to the respondent. However, there is one possibility to detect the incompleteness of the declared goals: if the goals are not described in all languages of the configurator, it is a clear sign that not all the goals are declared.

5.2.6.4 The Danger of Not Being Able to Express a Goal

Despite all the efforts to "draw out" his true goal from the stakeholder, sometimes we still have doubts that we have achieved the desired result. More often there are cases when the subject finds it difficult to express his desires or does not know what he wants. But the need to reveal his true purpose remains.

There are two possibilities to circumvent this difficulty.

The first is to create an environment in which the subject will not have to speak but to act purposefully, so his goal will manifest itself experimentally.

The explicit purposeful action of a person is an act of choice. If there are several opportunities in front of him, from which he needs to choose only one, he chooses the one that best suits his goal, even if it is not completely conscious.

Therefore, you can try to create a situation of choice. A catchy call is placed above the entrance to one American supermarket: "If you don't know what do you need, come to us, we have it!" After wandering through the floors of this huge store selling everything from fruits and toys to electronics and weapons, you will definitely see what you want to buy. So, our stakeholder needs to present a menu of intended, possible, and suitable goals that correspond to the situation in question.

However, in applied system analysis, another special variant of the situation of choice has been developed, when the subject expresses their wishes, they are not in the form of a statement of purpose, but in the form of a description of the final result. R. Akoff [4] called this method *idealized redesign*.

The next task is offered to the stakeholder. You don't like the existing system. Imagine that you can do everything without any limits! Abolish an existing system and design something that will satisfy you from scratch. Do not think about whether this can be done, if there are enough resources, etc. without any restrictions!

Stakeholders start work with curiosity. First, they break away from their narrow point of view and try to look at the situation as a whole for the first time which is interesting. Second, in the work, there is an element of the game "in the Lord God", — "I can do everything!". It is always interesting to play. But the most interesting thing is what comes next. When an idealized project design is checked for feasibility with available limited capabilities, it turns out that this will not be possible at 100%, but a fairly close approximation of it is possible.

An example is the approximation of the ideal and mass-unattainable goal of individual training for each student, that is, the creation of subgroups with more homogeneous compositions and different programs for them. (But be aware of the impossibility of changing a single element in a system: it requires to change many surrounding elements. In this example, you have to change the entire infrastructure of education.)

The second possibility of determining the true goal of the subject is to formulate it instead of him, as if to "calculate" the goal. This feature is based on the already noted tree structure of subjective goals. The subject cannot name the goal we are looking for, but it is on one of the branches of the tree. If you determine which branch, then it will be possible to find it, going down this branch. This task is facilitated by the fact that root goals and high-level goals, called life values, are fully realized by the subject, and he does not hide them, but on the contrary, he is proud of them. So, it is not difficult to find out the life values of a stakeholder. Publicists and scientists often pay attention to a certain opposition of such life values as "technocratic thinking" and "humanistic thinking" as different approaches to the formation of goals. Their main difference can be figuratively expressed by the slogans: "Man is the king of nature" and "Man is a part of nature". A more detailed opposition can be represented as Table 5.1.

These lists do not claim to be complete, as they only illustrate the difference between the two styles of thinking. Although such a comparison is made to emphasize (quite rightly) the dangers of a purely technocratic approach to the choice of goals, a complete rejection of all technocratic values would be excessive. For example, scientific and technological progress is not an alternative to social development, but its means; education can be considered as the antipode of culture only in isolation from it.

How different goals dictate these different value systems can be illustrated by a story of an American city. The municipality had to instruct the traffic police what measures to take to reduce the number of traffic accidents. There are two ways:

TABLE 5.1
Technocratic and Humanistic Values Systems

Technocratic	Humanistic
Nature is an unlimited source of resources	Natural resources are limited
Superiority over nature	Harmony with nature
Nature is hostile or neutral	Nature is friendly
Nature should be conquered	Nature in a delicate balance
Information technology development society	Sociocultural development of society
Market relations	Public interest
Risk and win	Security guarantees
Individual self-support	Collectivist organization
Reasonableness of means	Reasonable goals
Information, memorization	Knowledge, understanding
Education	Culture
Man is the means, the cog of society	Man is the basis and the purpose of society

hidden and open patrols. In the first, a police officer detects violators from an ambush and fines them; in the second, he openly demonstrates his presence on the road, and, seeing him, drivers follow the rules. The votes in the municipal council are divided. Some justified their decision with a desire to replenish the city budget with fines and restore order. Others pointed to the unethical, provocative ambushes, the obvious presence of a policeman that prompted caution to drivers.

The difficulty of the method of "calculating" a goal through the life value is that it is not one in a person (e.g., human interests put in the first place, others are the interests of the nation, third are of their company, fourth are families, fifth are personal, spiritual values, material, political, aesthetic, etc.). In different situations, a person can rely on different values. Which of them to try to find out in our case?

If this cannot be determined from the nature of the problem itself, then the answer lies in the configurator. After all, life values are expressed in some languages. Therefore, it is necessary to deal only with those values that are described in the languages of the configurator of the given stakeholder.

5.2.6.5 Peculiarities of Identifying the Goals of the Organization

The reasons for the discrepancy between the declared and true goals were discussed above when the subject ("physical person") unintentionally — by mistake or ignorance — declares a nontrue goal or not all true goals. The case of identifying the goals of the organization has to be allocated separately.

The fact is that the spokesman of the organization's goals is its leadership. And it has to announce the goals of the organization if there are conflicts between the interests of his system and the environment, as well as the interests of its system and its own interests. As a result, the goals proclaimed by the corporation do not always coincide with the true goals: there is often a big difference between what is being preached and what is being done in practice (which the system analyst should reveal).

In the first direction (management behavior in the event of a conflict of interests between an organization and society), R. Ackoff [4] gives such examples: "Many organizations proclaim care for the quality of the environment, but protect and support it no more than they are forced by the laws and public opinion". (Suffice to mention the behavior of oil companies in the virgin territories of the Russian North.) "Some companies that promote unlimited competition and a free market system turn to the government as soon as they feel that it infringes their interests. Others claim equal employment opportunities and carefully avoid employing members of national minorities and women".

The second direction is connected with the fact that the leadership of the organization inevitably has its own goals, which are sometimes realized at the expense of the resources of the organization. In such cases, the leaders simply remain silent about their goals, although in fact they are trying to realize them fully. As R. Ackoff writes, "many company executives claim that their main goal is profit maximization. However, an impartial check of their behavior reveals that this goal is not their dominant one. Otherwise, the directors would work in less luxurious cabinets, fly on scheduled flights, stay in medium-sized hotels, etc."

It is clear that most managers tend to donate at least part of the profits in the name of ensuring an acceptable quality of business life. Again, the point is not that such a goal is wrong or immoral; on the contrary, it should be extended to all workers. The desire of management to ensure a high quality of life is not a secret for many people who do not benefit from this desire, and they are indignant about this. Their simplified morality inhibits the development of corporations. Sometimes managers experience guilt in such cases. The stronger this feeling, the deeper they go into defense; therefore, the more they resist change. It also hinders development".

According to R. Ackoff, it is useful to divide goals at different levels of the tree into tasks, goals, and ideals. With the technology "top-down" first of all try to formulate the goal of the highest level — the mission of the organization. The mission is a common goal for all members of the organization, defining and uniting all the roles of the system in the environment and harmonizing the actions within the entire system. The difficulty in defining a mission is that the leadership of an organization is inclined to consider the mission of the goal of the system itself ("to become more profitable, to become larger (grow), to become better in the industry", etc.). A well-defined mission should reflect the interests of all stakeholders.

Gradation of goals can be defined in different ways. As an initial classification, R. Ackoff proposes to distinguish between:

1. *tasks* — results that are supposed to be obtained within the planned period;
2. *goals* — results that are not expected to be achieved beyond the planned period, but to which we expect to come closer within this period;
3. *ideals* — results that are considered unattainable, but approximation to which is desirable and possible.

Thus, tasks can be considered as means of achieving goals, and goals as means of approaching ideals.

In the case of solving the problem of an organization, it is necessary to determine not what the organization does not want (this is expressed in the problematic mess), but what the organization wants: its ideals, goals, objectives, and "target mess".

A very effective method for this is the *idealized redesign* of the organization, a method developed by R. Ackoff's school. It consists in not dealing with the elimination of the deficiencies of the existing system, but immediately beginning to design the desired system.

To this end, it is proposed to proceed from the assumption that the existing system "disappeared last night", leaving only its environment. Further, it is planned to build a project of such a system that the stakeholders would like to have right now if they could create such a system. There are only three restrictions on the project:

1. the project should be technologically feasible: not any science fiction (e.g., one should not assume that the communication in the system will be telepathic, but it is possible to plan the use of satellite or fiberoptic channels);
2. the project does not have to be feasible, but must be compatible with the environment (inherently), that is, do not contradict existing laws and regulations;
3. the projected organization must be capable of continuous improvement, that is, learning and adapting to changes in internal and external circumstances.

5.2.6.6 Techniques of Work with Goals

Let us discuss several technical aspects of the execution of the target detection stage.

When formulating goals, it is recommended to strive for clear, understandable definitions. Of course, it is not always possible to achieve measurable concreteness (e.g., in the case of the goal "To increase the morale of a sports team"), but we should strive to do this in every way ("By the end of the year, to reach such a production volume of such a product"). Experts recommend to try to achieve goals that are:

a. realistic, that is, achievable with cash financial, material, and time resources;
b. specific, so that any progress toward the goal contributes to the solution of this particular problem, and not another problem;
c. measurable, that is, would allow to track the process of movement to the goal by measuring, the costs of which are within acceptable limits.

Sometimes goals can be formulated as a positive mirror image of a negative problem formulation. However, more often, one problem gives rise to several goals, especially if we strive for their specificity.

The next question is about structuring the target mess. Sometimes (especially when there are no constraints in the means) you can work with the target mess without arranging it. But it is often necessary to establish the priority, the order of work with different goals. A convenient means for this is establishing causal relationships between goals, which is expressed in their alignment in the form of a tree-like structure, the "tree of goals". If such relations really exist, the goal of the upper level cannot be achieved until the lower goals are realized, and the question of work organization is solved automatically.

There are two most common ways to build a tree of goals. If the source material is a target mess, then it can be built into a tree according to the method described in the section on problematic mess (the problem tree was built there).

In the cases when the final goal is of a project nature, that is, you can first formulate a "global goal" (this also requires serious efforts), then the goals of the lower levels can be obtained algorithmically using the decomposition algorithm described in the book by F.I. Peregudov and F.P. Tarasenko [5]. Often, goal trees are built on the basis of intuition and commonsense, but this is fraught with reduced quality results.

These two methods of constructing a tree of goals are sometimes called "bottom-up" and "top-down" technologies, respectively.

We also note the possibility of a situation where the links between the identified goals are not lined up in a tree but instead in a network structure. This should not confuse us since it will not become an obstacle for the subsequent identification of discrepancies between the problem and target messes.

Questions and Tasks

1. Why is it necessary to identify the goals of stakeholders to design an improving intervention?
2. List the main reasons for the discrepancy between the declared and true goals.
3. What is a "goal tree" and how can it be built?

5.2.7 Stage Seven. Definition of Criteria

Abstract

To choose the best alternative, we need to use a measure of their quality. Such measures are called ***the criteria****.*

In the course of solving the problem, it will be necessary to compare the proposed options to assess the degree of achievement of the goal or deviation from it, as well as to monitor the course of events. This is achieved by highlighting some features of the objects and processes. These features should be associated with the features of the objects or processes of interest to us and should be available for observation and measurement. Then, based on the results of measurements, we will be able to carry out the necessary control.

Such characteristics are called *criteria.* Each study (including our one) will require criteria. How many, what, and how to choose criteria? This stage is devoted to the answers to these questions.

First is the number of criteria. Obviously, the fewer criteria needed, the easier the comparison would be. Hence, it is desirable to minimize the number of criteria, and it would be good to reduce it to one. Sometimes it works well. For example, UNESCO, in allocating funds to help underdeveloped countries in the humanitarian sphere, decided to provide subsidies to the most lagging behind. But the "lagging behind" is an evaluative term, and criteria are required by which to determine who is the lagging behind. It is remarkable that in a number of areas UNESCO managed to effectively solve this problem with a single criterion. The state of health is generally assessed in terms of child mortality, and the state of education in terms of the percentage of illiterate people in the country. At one university in the United States, a teacher's supplement for the quality of teaching was made dependent on one criterion — the product of the number of students by hours in his classes. In Sweden, the effectiveness of the traffic police is measured by the number of deaths in road accidents. In all these cases, the single criteria proved to be workable.

Unfortunately, more often than not a single criterion fails to adequately assess the quality of the object in question. For example, the criterion of speed of arrival of firefighters does not characterize the fight against fires as it is not associated with a decrease in the number of fires. Spending per student does not measure the quality of schooling. The number of students per teacher does not fully characterize the quality of training at the university.

Therefore, you have to enter some more criteria, differently describing the object, and complement each other.

Here is an example of how to choose criteria.

A firm of systems analyst was ordered to analyze the problem of improving garbage collection in a big city. At the stage of forming the criteria, the following criteria were first proposed: the cost of garbage collection per apartment, the number of tons of garbage collected per person-hour, and the total amount of garbage collected. These criteria were rejected as being unrelated to the quality of work. Criteria such as the percentage of residential areas without diseases, the number of fires due to garbage burning, the number of complaints of residents about the accumulation of garbage, and the number of people being bitten by rats were considered more successful. This example illustrates the criteria requirement to be as consistent with the goal as possible.

Yet, what are the criteria and how many of them to choose? The answer will become obvious if we understand that *the criteria are quantitative models of qualitative goals*. In fact, the criteria formed in the future in some sense represent and replace the goals: optimization by criteria should provide maximum approximation to the goal. Of course, the criteria are not identical to the goal, but it is the similarity of the goal to its model. Determining the value of the criterion for an alternative is essentially a measure of its suitability as a means to an end.

Thus, it is clear that it is necessary to choose such criteria and so many of them that together they are an adequate model of the goal. (However, how to implement this recommendation will have to be decided separately in each case. It is not always fully possible. But do not despair: as the ancient saying goes, "you can walk a lot in shoes that are a little tight".) Consequently, we come to multicriteria tasks — not only because there are multipurpose tasks but also because one goal often has to be displayed by several criteria.

When choosing criteria, you can sometimes use the experience of previous work. For example, in the analysis and design of technical systems are usually used criteria such as financial (cost, profit, etc.), inventory (number of products, range, etc.), operational (efficiency, reliability, etc.), survivability (compatibility with existing systems, adaptability to the environment, the speed of obsolescence, safety, etc.), environmental friendliness, ergonomics and a number of others. Another advice is to introduce at least three criteria for each characteristic described by the criteria: one should characterize the qualitative side, the other the quantitative side, and the third the temporal side. Such empirical lists are certainly useful and should be developed.

Criteria and limitations. Let us now pay attention to the fact that the set of criteria formed by us in the formulation of the optimality problem is divided into two subsets. Some criteria are subject to change as the situation approaches the desired state as close as possible. Others are subject to certain conditions, as a rule, fixing their values; these conditions must be met in the course of solving the entire problem. These criteria are called constraints. Recall (see Chapter 1, the notion of optimality) that constraints play no less a role in selection than maximized criteria. The difference between them is that the criteria maximized as if open up opportunities for the suggestion of more and more new alternatives in search of the best of them, and the restriction obviously reduces their number, prohibiting some of the alternatives. Some target criteria can be sacrificed for the sake of others, and the restriction cannot be excluded and must be strictly observed.

In the practice of systems analysis, there are cases when the imposed restrictions are so strong that they make it impossible to achieve the goal. Then the systems analyst must ask the decision-maker whether these restrictions can be relaxed or removed altogether. Recall the story of the bus company from Chapter 1. Another example is the imposition of strict requirements to the probability of false alarm, as presented to the developers of the radar station. This limitation would require unacceptably long periods of radar signal accumulation. Such a strict requirement stemmed from the reluctance to "too often" disturb superiors with false alarms.

Questions and Tasks

1. In what respect are the objectives and criteria?
2. What determines the set (number and nature) of the necessary criteria?
3. Discuss the similarities and differences between criteria and constraints.

5.2.8 Stage Eight. Experimental Study of Systems

Abstract

Often the missing information about the system can be obtained only from the system itself by conducting a specially planned experiment. The information contained in the experiment protocol is extracted, subjected to data processing, and converted into a form suitable for inclusion in the system model. The final action is model correction which includes the obtained information in the model. This section explains various measuring scales and correct processing measurements.

Experiment and model. Often the missing information about the system can be obtained only from the system itself by conducting a specially planned experiment. The information contained in the experiment protocol is extracted, subjected to data processing, and converted into a form suitable for inclusion in the system model.

The final action is the model correction which includes the obtained information in the model.

It is clear that an experiment is needed to improve the model. It is also important to understand that an experiment is impossible without a model as they follow the same cycle (Figure 5.9). However, rotation along this cycle does not resemble a rotating wheel, but a rolling snowball, with every turn it becomes more and more weighty.

Experiment and measurements. Various experiments can be described simply by their classification. If we do not interfere in the course of events but only register what happens at the inputs and outputs of the system of interest to us, then the experience is called a *passive experiment* (or *observation*). If we not only contemplate (and record) what is happening at the inputs and outputs but also affect some of them (some deliberately maintaining unchanged, while others changing properly), the experience is called *active* (or *controlled*) *experiment*. Like any classification, this only approximates reality. In absolutely pure form, these two experiments are impossible: active — because all the inputs and outputs cannot be controlled (some are even unknown), and passive — because every measurement and observation is an *interaction*, and it is impossible not to interfere with the result. The closest real, close to ideal, experiments are active laboratory experiments and passive observations in astronomy, history, archeology, psychology, etc.

Another important classification is dividing experiments into *direct* and *indirect*. A direct experiment is an observation of the characteristic that interests us (e.g., the gain of young animals can be measured by daily weighing). Sometimes the characteristic we are interested in cannot be directly measured, but there is an observable quantity associated with it; from the observations, we can extract the information we need; this will be an indirect observation (e.g., some actions of a mother can be judged on the strength of maternal love; those of prices on the cost; of blood pressure on the state of the cardiovascular system). The division of measurements into direct and indirect is important because they need to be treated differently, even if they are described using the same scale.

The results of the experiment are recorded in the form of a protocol of observations. This record is not the experiment itself but the description of its result, that is, its model. Understanding the term "language" widely, we can say that the protocol of observations is a record of the results of the experiment in some language. The variety of experiments is such that one language is indispensable; there are several such languages, called *measuring scales*. You should familiarize yourself with them because, in practice, you will

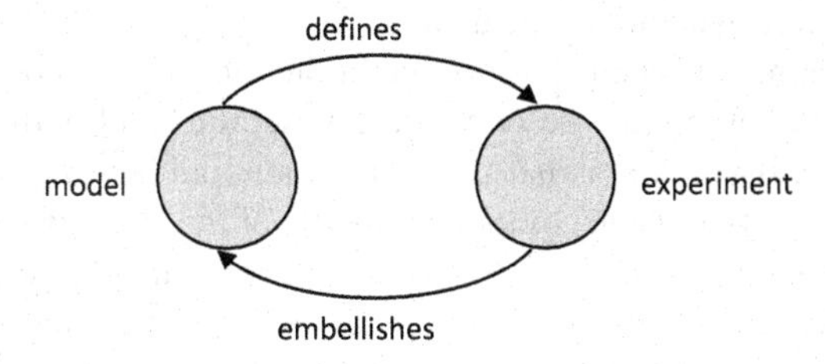

FIGURE 5.9 Experiment is a means for improving the model.

have to deal with data processing on different scales, and this should be done differently for each scale. As in any language, incorrectly constructed phrase loses its meaning, and incorrectly transformed experimental data do not carry the expected information.

Measuring scales. On the example of measuring scales can be traced the phenomenon characteristic of all languages (remember the discussion of language models in Chapter 3): starting with a universal but low-informative language, you can, including by attaching additional information to it, get more and more informative languages up to the most mathematized.

1. *Scale of designations (nominal, classification).* In the section on abstract models in Chapter 3, it was noted that the simplest model of diversity is classification. It is the basis of the *scale of names*. The measurement in this scale is to observe the classification characteristics of the object to determine to which class it belongs and to record it using a symbol denoting this class.

   Family names, diagnosis of diseases, number of houses and cars, players of sports teams, names of colors, addresses, etc. are examples of observations in the nominal scale.

   Let k classes be introduced: $A_1, A_2, \ldots, A_k$ (A_S is the name of s-th class). Let the objects $x_1, x_2, \ldots, x_N$ be successively observed (their set is called a *sample*, N is the *sample size*). With respect to each x_i ($i = 1 \ldots N$) is concluded to which of the classes $A_1, \ldots, A_k$ it refers ($x_i \in A_S$). The result is a protocol of observations.

   Since the only relationship defining the scale is the equivalence relation (the object either belongs to this class or not), the only admissible operation on the data in this scale is a match check, that is, a check for coincidence. This operation is represented by the special Kronecker symbol $\delta_{ij} = \{1: x_i = x_j; 0: x_i \neq x_j\}$, that is, the coincidence of two different observations (belonging to the same class) is denoted by one and the mismatch is zero. With the results of this *primary processing*, you can perform *secondary processing*. For example, the number of observations in one class with *xi* is $n_i = \sum_j \delta_{ij}$, and the relative frequency of this class will be n_i/N, the number of the most populated class $i_{max} = \{\arg(\max n_i)\}$; you can use different statistical procedures that use relative frequencies (e.g., the χ^2 test).

   It is possible to compare the data in the nominal scale obtained by different researchers only if they used the same division into classes (the number of classes and the boundaries between them must coincide). Only the names of classes and the order of their enumeration can differ as they do not violate the natural structure of the data.

   Once again, we recall that no operations on $\{x_i\}$, except for the comparison operation δ_{ij}, can be done; otherwise, the result will not be valid. For example, if you know that there are a hundred houses on the street, it is wrong to say that the geographical middle of the street is about the 50th.
2. *The preference scale (ordinal, rank).* If you enter an additional relation of the order between the classes of the nominal scale (preference; denote it by a symbol $\succ$), you get a new and enhanced information sense of the scale, called *ordinal* or *ranking*.

Examples of observations recorded in the ordinal scale include army and official ranks, school grades, magnitude of earthquakes (Richter scale), hardness of minerals (Mohs scale), wind strength (Beaufort scale), and prizes in competitions.

The valid transformation δ_{ij} (note at once that the valid transformations for weaker scales are also valid for stronger scales, but not vice versa) is now complemented by the preference check operation C_{ij}: $C_{ij} = \{1: x_i \succ x_j; 0: x_i \prec x_j\}$. Thus, the primary data processing in the ordinal scale consists of two valid transformations: δ_{ij} and C_{ij}. The use of other operations will lead to misunderstandings. An example is an unsuccessful attempt to account for the "average school score" when entering universities. In this failed experiment, the scores were treated as numbers and added up. The addition operation in the ordinal scale is invalid and gives a meaningless result. The experiment had to be canceled.

With the results of the primary processing (binary numbers), you can produce a suitable *secondary*. For example, you can define the number of observations x_i in an ordered series of all observations: $R_j = \Sigma_j^N C_{ij}$. This number is called the *rank* of the *i*-th object (hence, there is another name for this type of scale, *rank*). Other uses of the numbers δ_{ij} and C_{ij} are possible: in addition to finding frequencies and modes (as for the nominal scale), it is possible to determine the sample median (i.e., the observation with the rank closest to the number $N\backslash 2$), to divide the whole sample into parts in any proportion, to find the sample quantiles of any level p, $0 < p < 1$ (i.e., observations with the rank R_i, closest to the value of N_p), to determine the rank correlation coefficients between the two series of ordinal observations (r_S Spearman, τ Kendall), and to build other statistical procedures.

A few additional remarks on ordinal scales.

Various preferences are ordering in the presence of standard reference samples (e.g., the Mohs scale is based on ten specific minerals of different hardness), with fuzzy samples (wind force scale, school grades) in the absence of samples (sports, music competitions).

In addition to the scales *of perfect order*, unambiguously determining preferences (sequence numbering, military ranks, etc.), there are scales *of quasi-order*, when some elements of the ordered series are indistinguishable (mother = father $\succ$ son = daughter; uncle = aunt $\prec$ brother = sister), as well as scales *of partial order*, when there are incomparable pairs of classes (e.g., in sociological studies, the subject is sometimes unable to assess what he likes more — checkered socks or canned fruit, bicycle or tape-recorder, reading or swimming, etc.).

In ordinal scales, there is no concept of distance between classes, so any transformations that preserve order ("monotone") do not affect the information contained in the data. (Soldiers hanging on the shoulder straps a star and all higher ranks adding an asterisk will be prettier, but the essence does not change.)

3. *Scale of intervals (differences).* If the ordering of objects can be done so precisely that the distances between any two of them are known, then the

measurement will become noticeably more informative than that in the order scale. It is natural to express all distances in units, although arbitrary, but the same throughout the length of the scale. This means that objectively equal intervals are measured by the same segments of the scale wherever they are located (Figure 5.10).

As a result, it turns out that in our new scale, *scale of intervals*, the origin and the unit of interval length of the interval are arbitrary ($\Delta y \backslash \Delta x = \text{const}$).

Examples of observations recorded in the interval scale:

- Temperature (Celsius, Fahrenheit, Kelvin scales);
- chronology (from the birth of Christ, from Muhammad's move to Medina — 622 years later, from the Imperial dynasty in China — 5000 years earlier);
- altitude (from sea level; Holland almost all has a negative height).

The only new valid primary processing operation on the data in the new scale is subtraction, that is, determining the interval between two counts. For example, if we say that the temperature has doubled when heated from 9°C to 18°C, for those who are accustomed to using the Fahrenheit scale, this will sound very strange since on this scale the temperature will change from 48.2°F to 64.4°F. The division operation for this scale is also not valid. Only intervals make sense of real numbers. Using them (secondary processing), it is already possible to perform any arithmetic operations, as well as statistical and other procedures.

4. *Scale cyclic (periodic, differences).* There is a special kind of interval scale which is characterized by the fact that it is closed on itself, that is, after passing a certain period, its values are repeated. Examples include angular directions from a central point (compass scale, wind rose), time of day (clock face), phase of periodic oscillations (in degrees or radians), and geographic longitude (in degrees).

   Everything said about the interval scale applies to the cyclic one. To avoid misunderstandings, note that the addition of hours is not the addition of the timestamps themselves (which is an unacceptable operation), but the addition of *time intervals* between the origin and a clock, that is, secondary processing. We must also remember the conventionality of the start of the countdown (e.g., when switching to winter time, or crossing the date line). This scale is also called the *scale of differences* as it is invariant to shift to an interval, called the *period of the scale.*

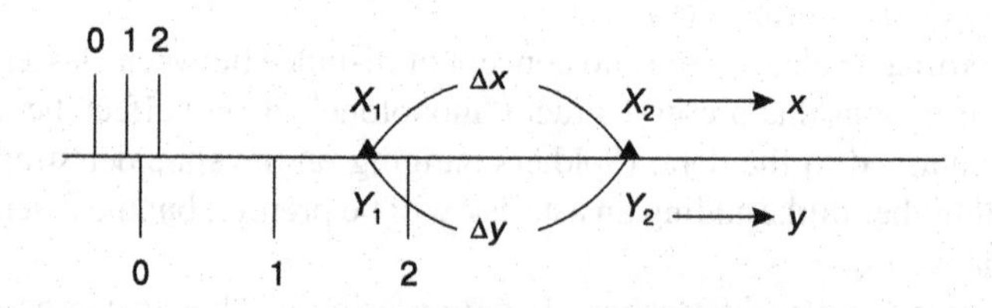

FIGURE 5.10 Interrelation between the various scales of intervals.

5. *Scale of ratios.* Introducing a defining relationship gives additional gain measurements. We demand that not only the ratio of the values of the same interval in different scales were constant, wherever this interval may be (that is typical for the scale of intervals), but that the ratio of the values of the same value measured in different scales were also constant, whatever place this value may occupy in reality (see Figure 5.11): $x_i/y_i = \text{const}, i = (1, \ldots N)$.

   The resulting scale is called the scale of ratios. Although the unit of measurement remains arbitrary, the zero mark becomes absolute and immovable.

   Examples of quantities whose nature corresponds to the scale of relations are:
   - length (cm, feet, yards, miles, etc.);
   - weight (kg, pounds, tons, etc.);
   - volume (m^3, barrels, liters, etc.);
   - money (rubles, dollars, euros, yen, etc.).

   Data in the scale of ratios become numbers to an even greater extent: in the primary processing, any arithmetic operations make sense, and the same can be done in the secondary processing.
6. *Absolute scale.* The previous "numerical" scales (of intervals and ratios) had degrees of freedom: intervals—two (arbitrary zero and unit), and relations—one (fixed, immovable zero and arbitrary unit). It is characteristic that the "numerical" capabilities of the data in these scales were limited: in the interval scale, only the operation of the difference, and in the scale of ratios, all arithmetic operations.

   Consider a scale that has both absolute zero and absolute unit. This scale has no degrees of freedom, and it is unique. These are the qualities of the numerical axis, which is naturally called the *absolute scale.* An important feature of the absolute scale is that the data values in it do not have dimensions, names, and its unit is absolute ("item"). This gives the data in this scale a special status (in English they are called *pure numbers*) — with them you can perform operations that are not valid with named numbers. (Record $2\,\text{cm}^2$ is valid, but $3^{2\text{cm}}$ is meaningless.) They can be used as an indicator of the degree, the base of the logarithm, and any trigonometric and other transcendental transformations are allowed for them.

   The numerical axis is used as a measuring scale when counting objects, and an auxiliary means is present in all other scales. The internal properties of the numerical axis, for all its apparent simplicity, are diverse and complex — the theory of numbers has not exhausted them to the end. Some dimensionless numerical relations, found both on the axis and in nature, cause amazement and admiration: prime numbers, Fibonacci numbers,

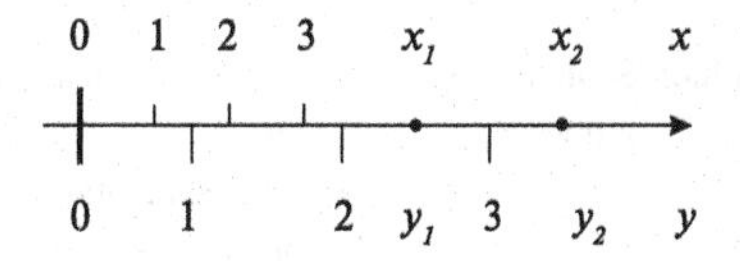

FIGURE 5.11 Interrelation between the different scales of ratios.

harmonic relations of sounds and sizes, laws of the theory of dimensions and similarity, quantum regularities, etc.

The final table of the basic measuring scales. The six measurement scales discussed above do not exhaust the diversity of languages in which we can talk about the diversity of reality. However, they are basic: the remaining scales are derived from them taking into account some third-party specific conditions. Table 5.2 summarizes some of the important features of the basic scales.

TABLE 5.2
The Final Table of the Basic Measuring Scales

Type of Scale	Defining Relations	Equivalent Transformations	Permissible Operations	Secondary Processing	Examples
Designations, nominal, classification	Equivalence: =	Renaming and reordering	Kronecker symbol $\delta_{ij} = \{1: x_i = x_j;\ 0: x_i \neq x_j\}$	Calculation of relative frequencies and their processing	Names, numbers of cars and houses, symbols
Reference, ordinal, rank	Same plus preference: $=, \succ$	Not changing the order, "monotonic"	Same (δ_{ij}), and comparison function: $C_{ij} = \{1: x_i \succ x_j;\ 0: x_i \prec x_j\}$	Same, and calculation of ranks: $R_j = \Sigma_j^N C_{ij}$, quantiles	Orderings, school marks, enumeration
Intervals, differences	Same, and constancy of intervals ratio: $=, \succ$, $\Delta y \backslash \Delta x = =\text{const}$	$Y = ax + b$, $a > 0, b \in R$	Same, and calculation of intervals: δ_{ij}, C_{ij}, and $\Delta_{ij} = x_{ij} - x_j$	Arithmetic operations over intervals	Temperature, chronology, height of locality, latitude
Differences cyclic, periodic	Same, and periodicity: $=, \succ$, $\Delta y \backslash \Delta x = =\text{const}$; and $y = x + b + n\tau$	Same, and periodicity	Same as above	Same as above	Clock, compass, seasons, phase of oscillation
Ratios	Same, and constancy of ratio of measurements: $=, \succ$, $\Delta y \backslash \Delta x = =\text{const}$, and $y/x = \text{const}$	$y = ax + 0$	All arithmetic operations	All arithmetic operations	Length, weight, volume, mass, area, money
Absolut	Same, and absolute zero and unit	Scale is unique: numerical axis	Any arithmetical and transcendental operations	Any operations on numbers	Calculation of something, of proportions, basis of other scales

Other scales. Measurement practice in various activities has led to the desirability of introducing scales that are different from the basic ones. Considering the purposefulness of this text to professionals of any specialty, we restrict ourselves to a brief listing of the most commonly used modifications of the measuring scales, so that if you wish you can find more detailed information.

Measurements of continuous quantities are very common, and their values are inevitably recorded with finite accuracy and rounded. A special case is represented by scales when a finite number of digits is determined not by the digit capacity of the recording device, but by the accuracy class of the measuring device, when it is pointless to increase the number of digits. Both types of scales, as opposed to integer discrete scales, are called *discretized.* Data processing in discretized scales has a number of features.

Another practically important class of scales is *nonlinear.* The intervals of these scales do not meet the conditions of additivity, that is, the "price" of a unit division of such a scale depends on which part of the scale this division exists. Examples include quadratic, logarithmic, exponential scales, "probabilistic paper", and many nomograms.

Attempts are made to fill the gap between "weak" (nominal and ordinal) and "strong" (numerical) scales: the hyperorder scale and the Churchman–Ackoff scale.

In principle, each researcher can build their own measuring scale to better represent their results. It is very important to emphasize that each scale must be accompanied by a list of permissible primary processing operations that are specific to this scale.

So far, we have discussed scales based on a clear classification: an element either belonged to a class or did not. Real life has led to the need to consider cases where the requirement of strict equivalence is not met, that is, when an element can simultaneously belong to two or more classes. Two approaches have been developed to describe such situations.

The first is based on the theory of fuzzy (vague) sets. In this theory, class membership is described by *the membership function,* which characterizes the degree of confidence with which we attribute an object to a class. For example, to what extent does a 40-year-old belongs to the "young people" class, and to what extent to the "middle-aged"? In this theory, the measuring scale is the scale of values of the membership function.

The second approach is to consider that the probability distributions of the variable to be classified may overlap. It means that the measured value may belong to either one of the classes with corresponding probability. When deciding whether a variable belongs to one or another particular class, we dissect the range of values of the variable into distinct classes, resulting in error probabilities (Figure 5.12). Processing of random variables is done in mathematical statistics.

In carrying out the experiment, the information obtained depends on numerous factors:

a. how the experiment was organized, what values and in what order the managed variables were assigned;
b. what are the noises, errors, and distortions of the observed variables;

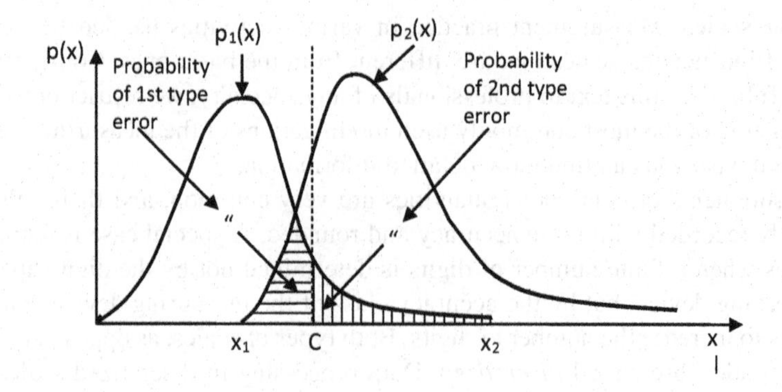

FIGURE 5.12 Statistical classification.

c. to what extent are fair the assumptions put in our model of the system under study;
d. what are the methods and algorithms for processing the experimental data.

The significance of these factors varies in different experiments; therefore, special theories have been developed on how to improve the quality of conclusions depending on a particular combination of factors. If necessary, you can refer to the appropriate ones:

– theory of interpolation and extrapolation;
– design of experiment;
– nonparametric statistics;
– robust statistics methods;
– optimization theory;
– search and strengthen patterns.

These keywords can be found in the catalogs of the necessary literature.

Why such details? Indeed, the proposed text is addressed to a wide range of readers: according to the author, knowledge of the basics of applied systems analysis is necessary, at least useful, for specialists of any profession. It is clear that if it comes to experiments, measurements, and processing of the data, especially in difficult conditions, it is rare for one person to perform all operation, and he will turn to specialists to get the final result.

However, practice shows that in trying to make it easier for themselves, experts often neglect some of the subtleties of data analysis, sincerely believing that they are insignificant. A typical example is the often used "digitization" of qualitative data—classes in ordinal and nominal scales are assigned numbers, and then these numbers are processed not as symbols, but as numbers, using arithmetic operations. But these are unacceptable operations for these scales! Another example is the weakening of data to bring it to monotony. In a table with disparate data, the strong scales are coarsened to the weakest (usually ordinal), so that the experiment protocol becomes single-scale, which facilitates processing. Unlike digitization, it is not the imposition of alien information, but the rejection of some useful information. This also reduces the quality of the conclusions.

The main task in the presentation of this stage is to direct the attention of the user of experimental services to check whether the use of invalid operations happened during the processing of data. For example, all numeric scales operate with numbers. But we already know that the numbers in the scale of intervals, ratios, and absolute should be treated *differently*. When accepting an order, it is recommended to check if there are any invalid operations in the processing algorithm.

Another important point to keep in mind is the coordination of the information power of the measuring scale with the information potential of the observed phenomenon. The stronger the scale, the greater the "information harvest" from the experiment. Therefore, each experimenter tries to use as strong a scale as possible. However, it is impossible to consider an observation with an arbitrarily set zero as belonging to the scale of ratios. Thus, when a *direct* observation is desirable, the scale of measurement makes you stronger, but not stronger than the very nature of the phenomenon.

The matter is even more complicated with *indirect* observations. The observed value, indirectly related to the unobservable phenomenon of interest to us, can belong to any scale, including the strongest scale; whereas the information potential of the phenomenon itself can be significantly lower. How to process such experimental data? The answer is that in the processing of indirect data, no matter how strong the scale they are recorded in, do not use operations that are unacceptable in the scale corresponding to the nature of the phenomenon under study. We will give one serious and two humorous examples.

Let us try to measure the power of a mother's love. This characteristic cannot be directly measured, but it is possible to record the number of rewards and punishments that a mother gives a child per day. The number of slaps and candies are fixed in the strongest, absolute (!) scale. The hypothesis is the assumption that these values are monotonously connected with the power of the mother's love for the child. But if one mother has the appropriate characteristics twice as better than the other, it is wrong to conclude that she loves her child twice as much; we can only say that she loves more because the power of love belongs to a qualitative, ordinal scale.

The second example is taken from medicine. The rate of precipitation when sodium citrate is added to a test tube containing blood is taken as an indicator of the intensity of the pathological process; the rate of precipitation is measured in millimeters per unit time. This idea is based on the fact that an increase in the intensity of inflammation leads to an increase in the content of globulin, which increases the rate of precipitation. The functional form of this relationship is unknown, and for different patients, it is different and nonlinear: the change in the proportions of sodium citrate or of deposition time leads to disproportionate changes in the elevation of sediment. Now, let for one patient medicine A led to a decrease in sediment (for 10 minutes) from 75 to 60 mm, and for another medicine B from 65 to 55 mm. Both the deposition time and the sediment height are measured in the scale of ratios. From here, it is impossible to conclude that the drug A is more effective since it led to a decrease in the sediment by 15 mm and the drug B only by 10! The intensity of inflammation belongs to the ordinal scale.

The third example is the test of mental abilities, which measures the time spent by the test subject to solve the test problem. In such experiments, although time is measured in a numerical scale, the measure of intelligence belongs to the ordinal scale.

Thus, the main pathos of the presentation of this stage of systems analysis is aimed at ensuring that, if necessary, the experimental study of the system with which you have to work, whether you will carry out experiments or will order them to others, would check if the experimental data are processed correctly. A typical situation is the widely used counting ratings case.

Questions and Tasks

1. What are "active" and "passive" experiments?
2. What is the difference between direct and indirect measurements? How should we take into account the difference between them when processing experimental data?
3. Have you learned the characteristics of the basic measurement scales? For example, have you noticed that quantitative scales differ in how you define the origin and units of measure?
4. Why should data from different measurement scales be treated equally always?
5. Try to identify the scales in which entries are made in each column on the page of a student's gradebook.

5.2.9 Stage Nine. Building and Improving Models

Abstract

Perhaps the most amazing thing in trying to understand how the world works is that taking into account only the finite sets of relations in an infinite world, we often succeed in achieving our goals. Whether the world is arranged "simply", either ourselves are very "limited", or our interaction with the world is "superficial" – these are philosophical questions, but the fact is that finite, simplified models allow us to successfully learn and transform (!) the infinite world. However, though not many models are suitable for this, some meet a number of requirements that we generalized in the concept of adequacy (see Chapter 3, Part I).

As already noted in Chapter 3, without modeling, no activity is possible at all. In the systems analysis, the problem situation model is needed to test possible interventions on it, to cut off not only those that are unimproved but also to choose the ones that improve the most (according to our criteria) among the ones that improve.

It should be emphasized that the contribution to the construction of a situation model is made at each previous and at all subsequent stages (both with one's own contribution and the decision to return to some early stage to replenish the model with information). Therefore, in reality, there is no separate, special "stage of model building". Nevertheless, it is worthwhile to focus on the features of building models, or rather "*completing* them" (i.e., joining new elements or removing unnecessary elements). Therefore, let's do this in the form of separation of these operations as if in a separate stage of analysis.

Much important for this work we have already discussed earlier.

First, in Chapter 2, we outlined the fact that there are only three types of models of a system: the *black box* (a list of essential connections with the environment), the *composition* (a list of essential compound parts), and the *structure* (a list of essential ties between parts), along with their necessary combinations. The difficulties faced by those who build these models were also discussed and will not be repeated.

Second, in Chapter 3, we discussed how to build models through *analysis* and *synthesis*.

Third, in Chapter 4, we examined the *method of trials and errors*, in which the completion, building, modification, and correction of models are carried out by including in it new information obtained during every next experiment with the system.

Fourth, in the section on language models, we emphasized that as the degree of knowledge of the system increases, the system model goes from its "soft", "loose" design in a verbal, qualitative form, through filling with new information (expressed in "professional" languages), up to (if necessary, i.e., if the problem has not been solved at the current stage) its increasingly "hard", more and more formalized description, in the end leading to a mathematical one.

Perhaps the most amazing thing in trying to understand how the world works is that taking into account only the finite sets of relations in an infinite world, we often succeed in achieving our goals. Whether the world is arranged "simply", either ourselves are very "limited", or our interaction with the world is "superficial" — these are philosophical questions, but the fact is that finite, simplified models allow us to successfully learn and transform (!) infinite world. However, though many models are suitable for this, some meet a number of requirements that we generalized in the concept of *adequacy* (see Chapter 3).

About qualitative models. Building "soft", "loose" qualitative models is more an art than a science. But the following are some useful tips:

1. It is necessary to divide all inputs of the task into controllable and unmanaged ones. Controlled variables are subject to us, and unmanaged variables characterize conditions and constraints of the problem situation.
2. When allocating controlled variables, it must be borne in mind that the correlative relationship between variables may be mistaken for causative ones. Let's consider an example. In one U.S. city, it was found that in areas where soot air pollution is greater, there is a higher incidence of tuberculosis. Effective measures were taken to combat soot emissions. After a few years, air pollution decreased significantly, but the incidence of tuberculosis did not. It turned out that malnutrition was the main cause of the disease. The connection with air pollution was indirect: in areas with a poor environment, the rent was lower and there were mostly poor families living that were poorly fed. So, the trap in this case is that we consider only known factors familiar to us from experience to be controlled. You can get around this danger by creating interdisciplinary teams of developers who look at the problem from different angles.
3. When considering uncontrollable factors, a very promising solution to a problem is the transformation of an unmanaged variable into a controlled

variable. (Let us recall the example of a bus company from Chapter 1 when it turned out that it was productive to bring conductors out of buses to bus stops at peak hours.) Of course, it's promising to study a factor that is not controlled due to lack of knowledge about it.

4. It is useful to keep in mind that the desire to bring all ties to cause-and-effect often leads to inadequacy of the model. The acorn is not the only factor leading to the growth of the oak, but many other conditions are needed, without which the oak will not grow out of an acorn: soil, moisture, temperature, light, etc. It is useful to use the concepts of directional correlation, producer–product, environment and conditions, etc.
5. Of the scientific constructions that significantly promote the construction of models in a "soft" situation, some examples include the theory of the situational management of the Moscow school, headed by D.A. Pospelov [6], and the theory of detection and amplification patterns laws of the Novosibirsk school of N.G. Zagoruiko and G.S. Lbov [7].

About quantitative models. In practice, increasing importance is attached to quantitative modeling. In this case, the "transparent box" model (a combination of black box models, the composition, and structure of the system) is embodied in the form of some formula or algorithm linking the input variables X with the output Y: $Y = f(X)$. Quantitative models can be descriptive and phenomenological when a formula is constructed heuristically and its coefficients are selected for best agreement with experimental data.

An important aspect of such a construction of quantitative models was highlighted by R. Ackoff in his 21st "f-law of management" [1]:

"The less managers understand their business, the more variables they need to explain it. $E = mc^2$ (special theory of relativity) contains one independent variable, m, and explains, perhaps, the most complex phenomenon understood by scientists. Then why does it take 35 variables to explain why people prefer a particular store and buy a certain cereal? The answer is obvious: these phenomena are not understood. The less clear something is the more variables are needed to create an alleged explanation for this.

That is why when managers do not understand what is happening, they collect all the information they can. Not knowing what information is relevant, they are afraid to miss something significant. As a result, they suffer much more from an excess of irrelevant information than from a lack of essential information".

Another form of quantitative model is preferred when the formula is derived from certain theoretical assumptions. In any case, the task is to identify the model, that is, determine the model parameters in which theoretical predictions and practical observations best fit.

Questions and Tasks

1. What is the difference between qualitative and quantitative models?
2. What does it mean to turn a black box into a transparent one?
3. What is "model identification"?

5.2.10 Stage Ten. Generating Alternatives

Abstract

In any system study, there comes a time when it is required to offer possible solutions to the problem. The process of advancing, inventing, and designing such options is called generating alternatives.

In any system study, there comes a time when it is required to offer possible solutions to the problem. In the described technology, this action is performed in two stages:

1. Identification of discrepancies between the problem and target messes. The differences between the present (and unsatisfactory) state of the organization and the future, the most desirable, ideal state to which one is supposed to aspire should be clearly stated. These differences are the gaps, the elimination of which should be planned;
2. the proposal of possible options to eliminate or reduce the detected discrepancies. The actions to be carried out, the procedures, the rules, the projects, the programs, and the policies are all the components of the management that should be developed.

The process of advancing, inventing, and devising such options is called *generating alternatives*. This is undoubtedly an act of creativity, and the question arises how to organize it and how to make it so that it is executed as best as possible.

Factors affecting creativity. Not being able to penetrate into the underlying mechanisms of the creative process, psychologists have, nevertheless, found numerous factors that influence the effectiveness of attempts to create. Both positive factors contribute to creativity and negative factors hinder it (do you recognize the black box model? See Figure 5.13). It is clear that in a conscious organization of the stage of generating alternatives, positive factors should be deliberately encouraged and used, and negative ones should be blocked, excluded, and muted. Since this can be done in different ways and in different combinations, there are many proposed and operated methods for generating alternatives. For example, in "hard" methodologies aimed at solving well-formalized problems (such as technical ones), there are dozens of such technologies.

For "soft" technologies that work with "loose" and poorly formalized problems, especially in the management of social systems, there are fewer well-established technologies for generating alternatives (see, for example, [4]). Before describing these technologies, let us describe the factors used in them.

The first group of factors includes the external conditions associated with the physiological characteristics of a person: temperature, lighting, air condition, sound background, and cozy environment, all of which affect the creativity. Without going into details, we will just recommend *to take into account these factors*: it is necessary to create some sufficient comfort for participants while generating alternatives, and suitability on the premises for work: availability of means for presenting results, computer support for work, communication facilities, etc.

The second group of "internal" factors is connected with our psychological features.

Among the positive factors, the most powerful for generating new ideas is *communication with other people* (it's not for nothing that scientists attach great importance to their participation in symposia, conferences; politicians to congresses, rallies; leaders to meetings; doctors to consultations, etc.). Hence, the recommendation *to conduct this stage in the form of collective, group work.* It turns out that people generate more ideas when interacting with each other rather than trying to come up with something separately. It has been proved that a group of experts who have information about a problem generates many more solutions to it, if they work and exchange information in the course of working rather than acting separately.

Of course, individuals differ in their creative potential, and it is desirable to have the highest possible result, gathering talented people, but in real situations, a number of factors limit us.

First, access to reputable, experienced, highly qualified experts is usually limited (including financial reasons); second, the requirement to take into account the opinions and interests of the direct participants in the problem situation under consideration (the goal is to create the improving intervention) implies that they must be involved in personal participation in a system study, so you have to work with those who are, with their various inherent creative abilities. Therefore, the technology and method of generating alternatives must be invariant to the inventive abilities of its participants.

The observation that "one mind is good, and two is better", has two bases. The first is that new ideas are an emergent result of combining several well-known ideas, so that a person searches the answer to a question rummaging through the nooks of his memory, looking for something that is useful. Since the information area (the world of models) is different for each individual, there is necessarily something that one of them knows, while the other does not. Therefore, the general ability to search for "suitable" is wider for two more than for one (see Figure 5.14); and, therefore, another recommendation arises by induction: the more people, the wider the general area of a priori information on which they will look for suitable ideas.

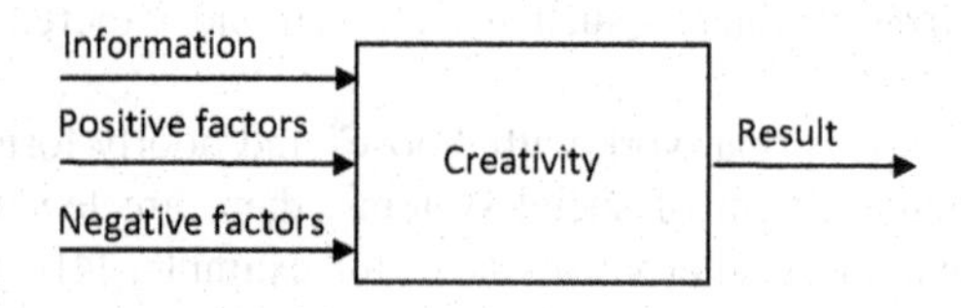

FIGURE 5.13 Black box model for creativity.

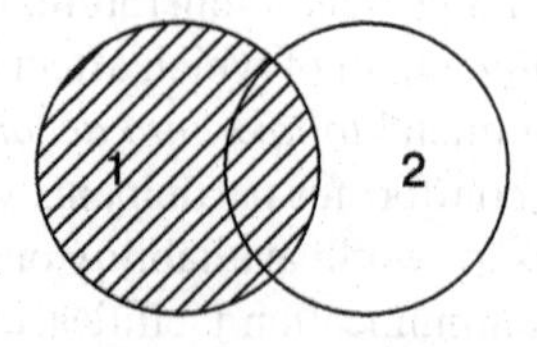

FIGURE 5.14 Two minds are better in size of the knowledgeable.

However, the emergent effect of group creativity is explained not just by the fact that individual knowledge is combined. A*ssociativity* begins to play the main role, which is a feature of our thinking. Left to itself, our brain is continuously working, moving from one thought to another. At the same time, each subsequent thought is connected with the previous one, as if repelling from it. As a result, the course of our thoughts can be depicted in the form of a continuous walk curve along the field of our information (Figure 5.15; the asterisks represent ideas "suitable" for our problem). Thinking alone, we use internal associations and look for useful ideas. If we hear what others have invented, then we sort of jump from our internal trajectory into an area of our memory where we wouldn't have looked at without this external influence. Thus, an external association is turned on, which may lead to a new idea that was not found when searching on internal associations (Figure 5.16). This is the second and main reason for the effectiveness of group creativity. Hence, the next recommendation: *to increase the efficiency of the technology for generating alternatives, it should ensure communication between performers.*

However, other psychological characteristics of people also come into play. It turns out that the *hard work on inventing solutions to a problem may not last long*: 30–45 minutes, a maximum of an hour (it's not for nothing that the academic hour in our educational institutions lasts 45 minutes). After this, the activity of the subject drops sharply, and communication between two people takes some time. This means that during the time of creative activity, a limited number of contacts (about 100) can be provided. This leads to a new recommendation: *the number of participants should not exceed 6–12 people.* (This restriction is not rigid, but deviation from it, less than six reduces and more than 12, does not increase the productivity of work.)

Together with the first feature of group thinking (combining individual knowledge), we can formulate another recommendation: *it is desirable to select participants for*

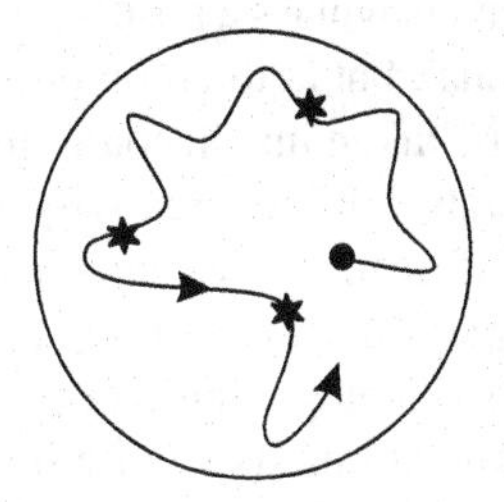

FIGURE 5.15 Trajectory of looking for useful information.

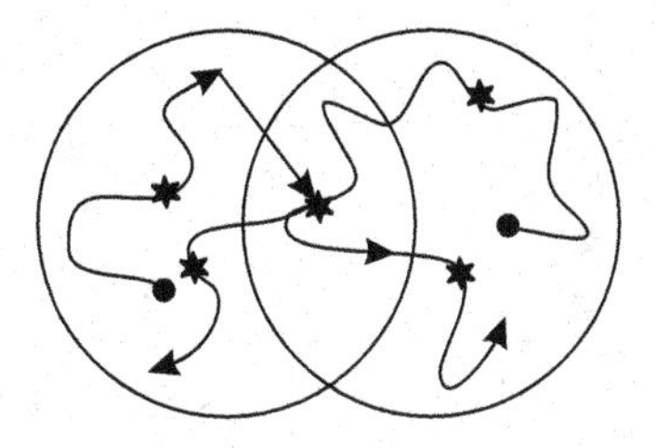

FIGURE 5.16 Trajectories of mutual looking for useful information.

generating alternatives as different as possible, according to their background, age, profession, education, life experience, etc. This will ensure the maximum expansion of the information area for the search of new ideas with a limited number of participants (see Figure 5.17).

Unfortunately, it can be noted that there are many more factors that negatively affect the creative process than positive ones. We have to take measures to neutralize them. Three such factors are noticeably distinguished by their strength. The first is the *responsibility for the proposal*. If a person feels that a proposal of an idea may be followed in some form by responsibility for it, he will most likely refrain from saying it. The advancement of an idea and its realization require different personal qualities, and they are rarely combined in one person. This is not always understood and accepted (recall the times when all scientists in the USSR were required to "introduce" their discoveries into practice, and the saying "The initiative should be punished", meaning the imposition of the execution of an idea on its author). Hence, the recommendation: *to release the participants in generating alternatives from responsibility for the proposals made*. This can be done in different ways: either legally (just as staff officers offering options for the upcoming battle are not responsible for the consequences of the option chosen by the commander), or organizationally, ensuring the anonymity of the author (following the example of scientific journals, hiding the names of referents who criticize submitted articles).

The second factor that seriously interferes with creativity is *criticism*. The human psyche is designed so that if they anticipate criticism in their personal address for the proposed proposal, they will not speak out at all. Hence, the following recommendation: *in alternatives generation technologies, measures should be taken to block personal criticism*. In some social systems, this moment is interpreted differently; let us recall the slogan of the Soviet times: "Criticism and self-criticism are the driving force of our society". In fact, criticism plays a positive role only if it is constructive, that is, it is not aimed at the individual, but at the promotion of competing ideas and their comparison. This should also be taken into account in the technologies of this stage.

The third reason that strongly inhibits creativity is the *a priori restrictions on the desired solutions*. In an effort to reduce intellectual effort, we consciously or subconsciously cut off those parts of our memory where "it is not worth" to look. In addition, external influences (ideology, faith, or prejudice) may impose their own prohibitions. But sometimes (and at this stage, often) it turns out that the most effective solutions lie precisely "where you shouldn't look".

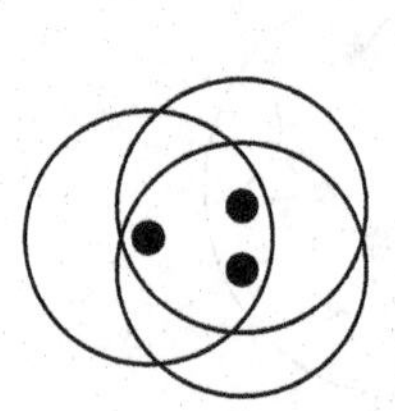
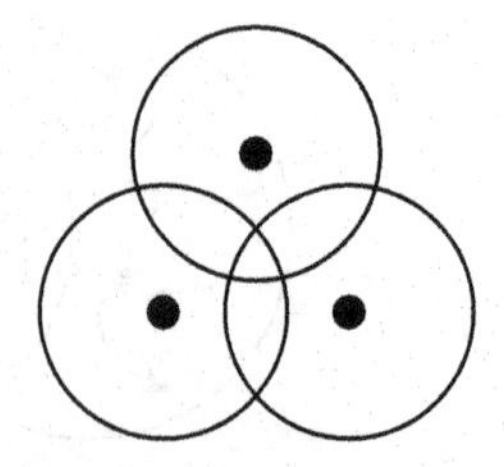

FIGURE 5.17 Equal number of subjects may have different competencies.

As a clear example, let us reproduce the problem of R. Ackoff [2]. Let us consider nine points on the sides and in the center of the square (Figure 5.18). Task: (1) connect all points with a continuous line; (2) the line must be broken (i.e., consist of straight line segments); (3) the number of segments should not exceed four. It usually turns out that many have difficulty in finding a solution, and only because they themselves have imposed a limitation: the broken line should not go beyond the square. With this restriction, there is no solution! This limitation is not in the formulation of the problem! It is necessary to refuse it, and as many as four solutions appear (one is shown in Figure 5.19). Some may be surprised to know that there are solutions with a number of segments less than four. But the surprise will disappear if you again remove the unconscious limitation that the sheet on which the points are drawn cannot be folded: it can be folded so that three, two, and even one segment will connect all the points. Moreover, if you remove a priori restriction that the line must be thin, then with the line of the whole square width you may solve the task without folding the paper with only one move.

Thus, we come to the recommendation that alternative technologies should include measures that, if possible, *remove or weaken a priori restrictions.* This is not easy since these restrictions are implicit. But in two examples you can see how this is done. Architecture students are encouraged to strain their imagination and design a house, provided that the force of gravity is not directed downwards but sideways. Sometimes the proposed new designs can be used in normal conditions. The second example is idealized design according to R.L. Ackoff [4] when performers are invited to create a project that suits them (instead of existing and nonsatisfactory) without taking into account its physical and financial feasibility. Subsequently, it

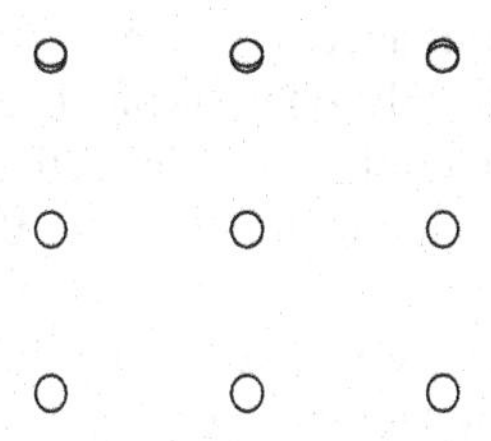

FIGURE 5.18 A configuration for setting a problem.

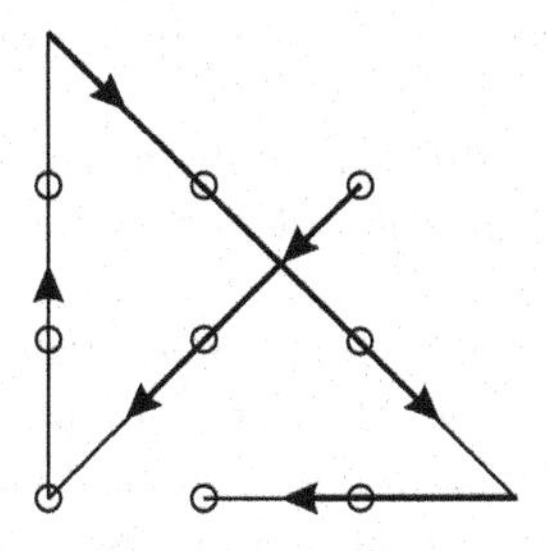

FIGURE 5.19 One of four solutions to the problem.

turns out that many of the qualities of an idealized project can be realized even with inevitable limitations.

For solving problems of managing an organization (management), recommendations can be made about how to remove a priori constraints. In this case, these limitations lie in which model of the organization the manager is guided by. The removal of restrictions consists of reporting about the existence of other models and proposing to look at their problem situation from the standpoint of those other models (see the fourth stage).

Different technologies for generating alternatives. Since the abovementioned (and also other) factors affecting creativity can be used in different ways and in different combinations, it is not surprising that many different technologies for generating alternatives have been proposed and are being used. There are different views on how to do it. In two points (comfort and collectivity), everyone agrees, and then the differences begin.

There are two fundamentally different approaches. Supporters of the former consider it absolutely necessary to separate the stage of proposing alternatives and the stage of evaluating suggested proposals. This approach is justified by the consideration that the evaluation of the alternative is its criticism, which can keep from offering other ideas. With this approach, the main goal of the generation of alternatives is to generate the maximum number of ideas, since this increases the likelihood of a really good proposal (the analogy can be artillery firing on the area where the target with unknown coordinates is located: the more shells are fired, the higher probability of hitting the target). Typical technologies of the first approach are brainstorming, Delphi method, and morphological analysis.

Proponents of the second approach proceed from the admissibility of criticism and discussion while generating alternatives and considering the goal is not a large number of alternatives, but access to a small but qualitative number of them (two or three, and sometimes only one). Typical technologies of this approach: TKJ method, synectics, search conference, dialectical approach, idealized design, Altshuler's TRIZ, devil's advocate technique, etc. Briefly consider some of these technologies.

Brainstorming. This a very effective, easily implemented, and widely used method proposed by A. Osborne "for the sole purpose of obtaining a complete list of ideas". An important feature of the method is an absolute ban on criticizing the ideas expressed; the participants are strongly warned about this and in case of violation of the ban, the guilty person is immediately removed from the group.

The brainstorming procedure can be described as the following sequence:

1. selecting several possible different participants;
2. familiarizing them with the rules of work, especially with the very strictly forbidden criticism;
3. presenting a summary of all available information about the problem;
4. using any psychological methods to increase the desire to achieve success, creating an atmosphere of working excitement and group inspiration;
5. explicitly formulating foreseeable a priori limitations and orienting the participants to overcome them;

6. using internal associations in which participants are invited to write down each of their idea on a separate card (for the convenience of subsequent analysis);
7. including external associations, each participant (in no particular order) announces its results of the previous stage. The remaining write down their new ideas emerging during the listening process;
8. announcing new cards continues until the imagination of the participants in the brainstorming is exhausted. This usually happens in 30–40 minutes with a score of several proposed alternatives. Then all the cards are collected and transferred for analysis to another group of experts (although quite often the participants in the brainstorming show a desire to participate in analysis as well).

This classic version of brainstorming is complemented by several modifications:

- before a "real" attack, carry out a "warm-up" brainstorming on a simple problem that is not related to the main one. This is especially useful if everyone in a group participates in the brainstorming for the first time;
- repeatedly alternate the internal and external associations: three to five minutes of silence (and reflections) are replaced by three to five minutes of discussing ideas, followed again by a period of silence, etc.;
- repeated brainstorming of the same problem in two or three days. Here we use the fact that in the subconscious, information is constantly being actively processed, and over the days following the attack, new ideas may arise; there are even known cases when ideas were born in a dream, on a walk, etc.;
- "reverse" brainstorming, in which the goal is not the generation of new ideas but the search for flaws in the existing ideas and the ways to eliminate them;
- written versions of brainstorming ("*brainwriting*"):
  a. method "6-3-5". Six people write down (each on their own sheet) three ideas. Then the sheets are transferred to the neighbor on the left who complements the ideas of the leaf that has come to him by his own associations in the development of the original three ideas. Then the next transfer of the sheets in a circle in the same direction is made, and the next participant replenishes the initial three ideas with his own additions. There will be five such transfers (hence the name of the method). The result of the work will be six sheets, on each of which there are three root ideas of the first author and their development by five coauthors. If each coauthor adds only one alternative to each of the three original ones, then even then there will be 18 ideas on each sheet and 108 on all six sheets,
  b. the "*brainwriting pool*" ("written pooling") method. Unlike "6-3-5", there can be any number of participants. Leaflets with the first three ideas are thrown into the general heap ("pool"), from which each one at random takes another sheet, complements it, and returns to the pool. (If you accidentally came across a piece of paper that you already had, it is replaced by another one.) The process is completed at the request of the participants.

The effectiveness of brainstorming is widely known.

The Delphi method is more often used for collective expertise, but it can also be used to generate alternatives. In this version, the method differs from brainstorming in that the participants are kept anonymous and their proposals are made public without attribution of authorship. This was first carried out by handing over the cards to the coordinator, who replicated them in printed form and distributed them to the participants; in recent years, software has been created that supports the Delphi procedure in the network configuration of personal computers: each participant enters his offers via the keyboard and receives the offers of all the others on the screen of his computer. Due to the anonymity of the texts, the Delphi method removes the ban on criticism: it cannot automatically have a personal focus. Therefore, there are peculiar dialogues and discussions entailing the introduction of amendments and additions to the original versions.

Although theoretically the number of cycles of the Delphi procedure is not limited, in practice, it ends after three or four iterations.

Morphological analysis. Another original simple and very effective method for generating alternatives was proposed by Zwicky. Participants are encouraged not to invent alternatives on their own, but only to propose their required properties and reasonable gradations of these requirements. This is really an analytical method, essentially, it is a matter of generating a composition model, that is, a set of components from which alternatives can be subsequently constructed. The scheme for morphological analysis is given in Table 5.3.

The participants try to propose as many requirements as possible for the future system, the appearance of which should solve the problem, and for each requirement to find a rational set of gradations for implementing a certain degree of expression of the proposed quality. For example, we will conduct a morphological analysis of the problem of choice in the section "Stage Eleven. Choice or Decision-Making".

Morphological analysis generates a large number of alternatives. After all, alternatives can differ by at least one value. Consequently, the number N of alternatives is $\Pi^m{}_j\ n_i$, where m is the number of qualities; n_j is the number of grades of the jth quality; N easily reaches many thousands for a single act of analysis. The care of the specialists will subsequently decide which of these alternatives deserve to be implemented; and participants in the morphological analysis have only one concern, that

TABLE 5.3
Morphological Analysis of Options

Properties	**C1**	**C2**	**C3**	**...**	**Ci**	**...**	**Cm**
Property gradations	C_{11}	C_{21}	C_{31}	...	$Ci1$	...	$Cm1$
	C_{12}	C_{22}	C_{32}	...	$Ci2$	...	$Cm2$
	...	...	...	...		...	...
	$C1j$	$C2j$	$C3j$	...		...	C_{mNm}
	...			...		...	
	C_{1N}			...		...	

is, to offer as many options as possible. The advantage of morphological analysis is the possibility of its implementation by one person.

We now turn to other technologies that generate alternatives but do not strictly separate the generation of alternatives from other systems analysis operations, both from preliminary steps (e.g., structuring a problem) and subsequent ones (e.g., evaluating and screening out the alternatives themselves).

The TKJ method (this abbreviation is not deciphered in the available sources) was proposed by S. Kobayashi and A. Kawakita. In this method, it is first proposed to find a general formulation of the problem situation acceptable to all stakeholders, and then formulate a generalized solution to the problem using the same technology.

The TKJ method includes the fact that the participants generate, at the first stage, the statements of facts relating to the problem, and at the second, proposals for concrete actions to solve it. Then the ideas submitted are aggregated and summarized to a single sentence.

The algorithm of the TKJ method can be summarized as follows:

Phase 1. Identify the problem.

1.1. Identify a concern.
1.2. Participants record the facts related to the problem, with each fact listed on a separate card. (The facts cited by experts must meet two conditions: (1) they must be substantially related to the problem; (2) they must be objectively verifiable.)
1.3. Cards are collected and multiplied by the number of participants, and each participant receives a complete set of cards.
1.4. The entries on each card are alternately read out loud.
1.5. Participants select from their own set of facts related to the fact just announced, forming a group of interrelated facts.
1.6. To each such group of facts the participant looks for a generic name that reflects the essence of the entire group. These names are entered into individual cards.
1.7. Then, Steps 1.4 and 1.5 are performed until all the facts are entered into a group.
1.8. Cards are announced with the names of groups.
1.9. Participants try to combine group names into new, more general groups with their generalizing characteristic descriptions. This process continues until a single description is obtained, covering all the facts. This will be a description of the problem situation (the problem mess) as a whole.

Phase 2. The search for improving intervention.

2.1. Brainstorming (see above).
2.2. Cards with suggestions for possible solutions to the problem are multiplied and distributed to participants.
2.3. The content of the next card is announced.

2.4. Participants select offers that can be combined into a single, more general offer.
2.5. This summary sentence is formulated explicitly and recorded on a new card of the next level.
2.6. The generalizations are announced again, and efforts are being made to combine them into even more general formulations until the most general formulation for the problem-solving measures is obtained.

As a result, a project of improving intervention should appear, taking into account the interests of all parties involved in the problem situation.

Synectics. Synectics is a procedure developed by W. Gordon, who defined it as "a combination of different and seemingly incompatible elements" to resolve a paradox or problem. It is based on thinking by analogy and generation of metaphors, that is, establishing the compatibility of completely different concepts in relation to one problem. The aim of synectics is to produce one but of high-quality alternative.

Synectics technology begins with defining the "essence" of the problem situation. Its conflicts and paradoxes are expressed in the form of a question, covering in its most general form the uniqueness of the situation, and then a metaphor is sought that answers it. The metaphor is sought in the area as far as possible from the problem situation; for example, if the problem is related to hearing, you can try to look for a metaphor associated with vision or touch.

The illustration can be a problem caused by the fact that one of the people living in one room wants to listen to loud music, while the other wants to read in silence. The essence of the situation is expressed by the question: "How to achieve loud silence?" The metaphor answering the question: "A bullet that has entered one body cannot hit another". This suggests a proposal for a lover of loud music to use headphones.

The procedure of synectics can be described as follows:

1. Search for the essence of the problem.
   1.1. Description of the problem.
   1.2. Analysis of the problem.
   1.3. The wording of the essence of the problem.
2. Search for metaphor.
   2.1. Transform the wording of the essence into the form of a question that evokes memories and emotions that activate the imagination.
   2.2. Finding the metaphor that answers the question.
   2.3. Analysis of the resulting analogy.
3. Search for a solution to the problem.
   3.1. Determination of functional requirements for the solution by "fitting" the analogy to the problem.
   3.2. Final wording of the decision.

Search conference. F. Emery and T. Williams have developed a "search conference" method, that is, a research seminar whose goal is to find effective ways to adapt the organization to environmental changes. The method is close to the well-known SWOT analysis in economics and management, but differs in details. Algorithmically, search conference can be described as follows:

1. participants are encouraged to give their perception of "trends in society as a whole";
2. the answers are summarized to create a "picture of changes taking place in the broad social field to which the system in question belongs, but over which it does not have (almost or completely) direct control";
3. participants consider "the forces that determined or are likely to determine the evolution of their organization or community. At this stage, participants can make important judgments about the respective goals of their system";
4. they should "identify the limitations that are inevitable stem from a lack of resources, from existing structures and culture";
5. they further formulate "strategies for planned adaptation";
6. then "the question of what steps are needed to begin the agreed changes" is discussed.

The dialectical approach to solving problems was developed by S. Churchman, R. Mason, J. Emshoff, I. Mitroff, and others. The main goal is to clearly highlight the assumptions on which the problem-solving plan will be based. The method consists in the deliberate collision of contrary opinions. Two or more teams deliberately develop different and conflicting solutions to the same problem. Groups are completed so that they represent different positions (different levels of hierarchy in the organization, short- and long-term planning horizons, different strategies, etc.). Confrontation between them and their decisions exposes the assumptions underlying their proposals. During the debate, parties are allowed to ask questions to clarify and elaborate positions. For example, what is the difference between their initial assumptions? Which of the stakeholders is the most important for the party? How different are their assessments of the importance of the intended interventions? What assumptions of the other side cause the greatest doubts or objections? Following the discussion, the parties may make changes to their proposals.

Then the decision-maker synthesizes his decision taking into account the arguments of both parties.

As in most situations of rivalry, the dialectical approach is dramatic. Each of the parties in the final debate tries to do everything possible to convince the decision-maker that its proposed option is better than the other. The arbitrator, after hearing both conflicting parties and using the arguments of both, makes his own decision.

The dialectic procedure exposes the arbitrariness of some restrictive assumptions and reveals their consequences. It focuses on the implications of the data and shows that the same data can be interpreted differently depending on the assumptions made.

You can set forth a dialectical process in the form of the following algorithm.

1. Preparation.
   1.1. The decision-maker creates two or more groups, providing them with identical targets and a set of source data.
   1.2. Each group develops a solution that is wittingly in conflict with the proposal of the other group. In this case, the assumptions are formulated as explicitly and clearly as possible.
2. Confrontation.
   2.1. Each team presents its version to the decision-maker in the presence of the opposing side and protects it with all his might.
   2.2. After the presentations, each team attacks the other team, trying to weaken the positions of the enemy and strengthen their own.
   2.3. The decision-maker can ask questions of any of the parties at any time.
3. Synthesis.
   The decision-maker formulates his own decision using the information provided by both warring parties, and explicitly identifies the arguments on which the arbitration decision is based so that they can be followed during the implementation of the decision.

Idealized design. We have already described the method developed by Ackoff when considering the target detection stage. We now give a more detailed description of the method.

An idealized project is what stakeholders would like to have if they could have any desired system. This project should be technologically implementable and operationally viable, that is, could exist, being implemented, but it can be made without taking into account when and how it can be implemented. In addition, the system must be designed to be able to learn and adapt quickly and efficiently. This requires that:

1. system stakeholders could make changes to the project if desired at any stage;
2. if an objectively unsolvable condition of the project now arises, then a practical procedure must be built into the system design itself to resolve it;
3. the system should be designed so that all decisions made in it, and the assumptions on which these decisions are based, are subject to continuous monitoring.

The result of such a design is not utopian and is not an idealistic system since it can be improved from the outside and self-improved; rather, it is the best system, striving for the ideal that its creators can imagine.

Idealized design can be described in the form of three actions:

1. the choice of mission — the general purpose of the designed system, its role in the encompassing larger system, of which the system itself and its stakeholders are a part;
2. determining the properties of the future system desired by the designer and stakeholders;
3. system design, that is, determining how to implement named properties.

The final project should cover every aspect of the system: its social and technological processes, organization, management system, inputs, outputs, etc., that is, all the three basic models of the system (black box, parts, and structure).

Two versions of a project are usually made: limited (i.e., without significant changes in the environment of the existing system) and unrestricted (i.e., with its change).

The process of idealization frees its participants from even the unconsciously imposed restrictions on themselves since it discards feasibility considerations. When such a project is completed, its developers usually realize that most of its parts are completely realizable and that they themselves are the main obstacle for the future they desire.

Interactive planning consists of six interrelated actions that are carried out simultaneously since the entire process is carried out continuously:

1. the formulation of the problem mess;
2. planning goals (technologically feasible);
3. planning of means (policies, programs, projects, operations, etc.);
4. planning resources (money, personnel, equipment, premises, necessary goods and services, information);
5. designing the process of implementing plans (who, what, where, and when should do, how resources are allocated);
6. management design (how the organization will be monitored, trained, and adapted).

A detailed description of this method is given in the book by R. Ackoff et al. [4].

About other technologies. There are several other technologies that contribute to the creative process, which are not based on theoretical principles, but on practical experience; they are more "engineering" than "scientific". Just as engineers launched rockets earlier than ballistic theory was developed; as bridges, dams, and buildings were built before the creation of theories of structures and materials, and a number of procedures that improve the creative process were proposed without defining the concept of creativity itself and the theory explaining it. A section devoted to the review of such methods is contained in the article by R. Ackoff and E. Vergar [7] (it should be immediately noted that a substantial part of this paragraph is based on this remarkable article). Among the methods mentioned there, we present the method of "lateral thinking" proposed by de Bono 30 years ago, which consists of restructuring data and conceptual combinations into something new.

Edward de Bono's approach is based on the idea of attracting right-hemisphere (imaginative, emotional, intuitive) thinking to generate alternatives, along with left-hemisphere (logical, formal). In most of his publications, he proposed a collection of heuristic, empirical techniques that promote the emergence of new ideas (e.g., the "Ten rules of simplicity" by E. de Bono [8]) or such tips as:

1. generate alternative ways to describe the situation;
2. explicitly formulate assumptions and subject them to critical examination;
3. identify duplicate topics and modify them;

4. identify the use of stereotypes and replace them;
5. identify repeated obstacles and overcome them;
6. consciously direct attention to areas previously unnoticed;
7. identify aspects of the situation, considered for the first time, and already known but viewed from a different viewpoint;
8. find other ways to decompose and aggregate the problem situation;
9. use the random introduction of new elements into the problem situation.

However, de Bono proposed two alternative generation technologies.

The first technology is named by him "Random Entry", a method of random associations. Randomly (e.g., at random from the dictionary) a few words are picked out, and then the group looks for associations of each word with a solvable problem with the aim of solving it. For example, to search for ideas on attracting bank customers by providing them with new services, the words "group (of musicians)", "church", "computer", and "joke" were chosen. Associations of the first word gave rise to proposals for background musical accompaniment in the bank; according to the second, to use pleasant aromas in the bank, as in a church; among the ideas for the third word, providing customers with free access to the Internet at bank branches, customer protection measures while using an ATM, donating electronic calculators to customers; on the fourth, the creation of a witty marketing slogan showing funny animated cartoons in the waiting room; etc.

The second technology of E. de Bono — "Six Thinking Hats" — alternates group consideration of important aspects of the desired solution to the problem. According to de Bono, the comprehensive consideration of any problem consists of taking into account six aspects. In a number of technologies (e.g., in ordinary brainstorming), each participant is invited to reflect in any of these areas at his discretion. De Bono believes that due to the impossibility of focusing attention on all aspects at once, many useful ideas are missing. He suggests focusing the group on each of the six main aspects in turn. To do this, put one of the six colorful hats on the table, and the group conducts brainstorming only in this direction:

1. collection of facts — white hat;
2. positive aspects: opportunities — yellow hat;
3. negative aspects: threats — black hat;
4. options that differ from the proposed — green hat;
5. the emotions associated with the proposed option — red hat;
6. organizational and operational aspects — blue hat.

As we see, the method is aimed not only at proposing alternatives but also at questions of their implementation and consideration of the consequences.

In conclusion, we note that all the above methods are designed to solve "soft" and "loose" problems ("hard" problems are solved by optimization methods). The following list of methods contains those that are most tested and made public; this does not exhaust all possibilities and there are many publications on others (mostly empirically found) technologies. Let us single out among them the Altshuler theory of inventive problem-solving (TRIZ).

Questions and Tasks

1. Why at this stage group creativity is preferable to individual creativity?
2. What are the three factors most hindering the creative process?
3. Describe the brainstorming algorithm.
4. Why the Delphi technology allows criticism of the ideas expressed?
5. How is morphological analysis different from brainstorming?

5.2.11 Stage Eleven. Choice or Decision-Making

Abstract

Sooner or later there comes a time when further actions may be different in means, leading to the same final result, with only one action being realized. There comes a moment of choice. Naturally, the desire arises to understand what a "good" choice is, to work out recommendations, how to get closer to the best solution, and if possible, to offer an exact algorithm for obtaining such a solution. It turned out that various situations range from well-studied, fairly formalized, mathematically described (so-called hard) situations to poorly structured, described in colloquial or professional, far from mathematical, languages (soft) situations, and with various mixed intermediate options.

Choice as an attempt to achieve the purpose. Sooner or later there comes a time when further actions may be different, leading to different results, with only one action being realized. It is already impossible to return to the situation that occurred at that moment. There comes a moment of choice.

Naturally, the option that most (in the opinion of the one who chooses) corresponds to his purpose is chosen. It is the choice that leads to the realization of the purposefulness of the entire activity of the subject.

The ability to make the right (i.e., the closest to the purpose) choice is a very valuable quality, inherent to people in varying degrees. Great commanders, outstanding politicians, brilliant engineers and scientists, talented administrators differ from their colleagues or rivals primarily in the ability to make better decisions and better choices.

Naturally, the desire arises to understand what a "good" choice is, to work out recommendations, how to get closer to the best solution, and if possible, to offer an exact algorithm for obtaining such a solution. It turned out that various situations range from well-studied, fairly formalized, mathematically described (so-called "rigid", "firm", *hard*) situations to poorly structured, described in colloquial or professional, far from mathematical, languages ("soft", "loose") situations, and with various mixed intermediate options.

For "hard" choice problems, a strict formal method of finding the best (optimal) solution under the given conditions is developed. In the case of "loose" problem-setting, the nonuniqueness of the solution is realized, and a "soft" technology is developed for finding acceptable, "improving" interventions. In intermediate cases, the person's intellectual abilities to solve informal problems and suitable formal methods of mathematics and computer modeling are combined in different proportions: decision support systems, expert systems, databases, automated control systems, etc.

At the previous stages of systems analysis, everything necessary for making the choice was prepared: a set of alternatives on which to make a choice (stage ten), the goals for which the choice is made (stage six), and the criteria for comparing the alternatives in terms of their suitability for achieving the goals (stage seven). The current stage is devoted to the consideration of the problems of the choice itself, that is, of decision-making process.

In the most general form, the choice can be defined as having a special purpose narrowing the set of alternatives: a part of this set X is considered acceptable ($C(X)$ in Figure 5.20), while the rest are rejected. Usually, they try to reduce $C(X)$ to a single alternative, but sometimes this is unwise or even impossible.

The desire to ensure that our choice was as correct as possible leads to the construction of a certain theory of choice that would suggest the means of synthesis of choice algorithms and their analysis (comparison). However, attempts to build a "general theory of decision-making" face serious difficulties.

The multiplicity of problems of choice. Let us find out various situations of choice using the method of morphological analysis. In accordance with this method, we list the factors that determine the nature of the choice and their gradations.

1. The set of alternatives X can be finite, countable, or continual (which requires different optimization methods).
2. Criteria type can belong to different measuring scales (we roughly break them down into qualitative and quantitative, but in case of necessity, we may use a proper set of measuring scales).
3. The number of criteria also influences the method of choice: the difference between single and multicriteria tasks is quite significant.
4. The number of decision-makers also leads to completely different ways of choosing (we will distinguish between one-sided and multilateral choices).
5. The degree of agreement between decision-makers significantly influences the method of choice. Decisions are made differently when the interests of the parties coincide (collective choice) and when they are opposite (choice in a conflict situation). Intermediate cases are possible (compromise choice, coalition choice, choice in conditions of conflict, choice with variable conflict).

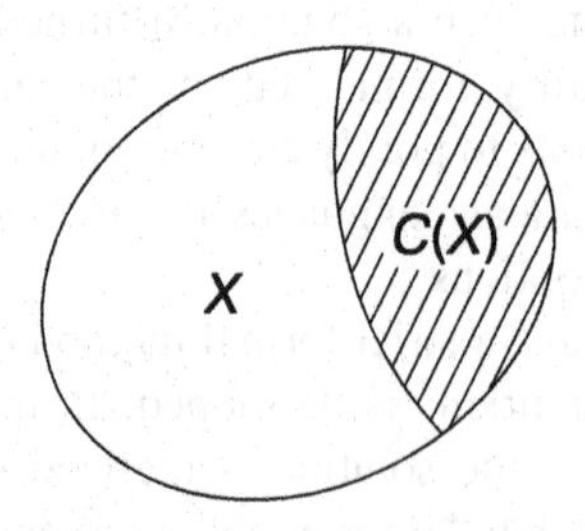

FIGURE 5.20 Choice as selection of acceptable variants.

6. The nature of the uncertainty of the consequences of choice varies from complete certainty (when the consequences of choosing each alternative are accurately known) to different types of uncertainty: ignorance of the consequences, knowledge of the probabilities of outcomes, and fuzzy uncertainty. Each of these options requires a completely specific approach and other mathematical methods.
7. Repeatability of the situation of choice. Different approaches to decision-making in a one-time (unique, nonrepeatable, first) choice and choice repeated, repeated in similar situations, allowing the use of previous experience, with or without a teacher, etc.
8. Responsibility for the consequences of choice. Wrong choice leads to losses. Losses can be acceptable, small, and can be intolerable or unacceptable. Of course, in these cases, the choice must be made differently.

Already taking into account only these factors gives 3·2·2·2·4·4·2·2 = 800 variants of the problems of choice. Each of them requires a special combination of methods from different fields of knowledge. It becomes clear why there is not (and cannot be) a universal theory of decision-making. Indeed, various theories have been developed for different types of choice situations: optimization theory (certainty of the outcome, one-sided choice, single or multicriterion problems); mathematical statistics (stochastic uncertainty); fuzzy sets theory (fuzzy uncertainty); collective choice theory (multilateral choice with unity of purpose); game theory (multisided conflict choice); etc. For some situations, algorithmic solutions have not yet been found (implicitly setting criteria, ignorance of essential parameters, etc.) when you have to act "by intuition", "according to common sense", "at random", etc.

5.2.11.1 An Overview of the Most Common Situations of Choice and the Decision-Making Methods Used in Various Cases

Criterial choice. The basis of this choice is the assumption that each individual alternative can be estimated numerically (by value of the criterial function). Then the comparison of alternatives is reduced to the comparison of the corresponding numbers.

Let x be some alternative from the set X. It is assumed that for all $x \in X$ a function $q(x)$ can be given, which is called a *criterion* (*quality criterion, conditional function, preference function, utility function*, etc.) and has the property that if the alternative x_i is preferable to the alternative x_j (we denote this by $x_i \succ x_j$), then $q(x_i) > q(x_j)$, and vice versa: $[x_i \succ x_j] \Leftrightarrow [q(x_j) > q(x_i)]$.

If we now make another important assumption that the choice of any alternative leads to unambiguously known consequences (i.e., the choice is carried out under conditions of complete certainty) and the value of $q(x)$ expresses this effect numerically, then the best alternative x^* is, naturally, the one that has the highest value of the criterion:

$$x^* = arg\left[maxq(x)\right]$$

$$x \in X$$

The task of finding x^*, which is simple in formulation, often turns out to be difficult to solve since the method for solving it (and the very possibility of solving it) is determined both by the character of the set X (the dimension of the vector x, and the belonging of its components to the finite, discrete, or continuous set) and the type of criterion (is $q(x)$ a function or a functional, which one, given explicitly or implicitly, in the form of equality or inequality, etc.). University course of optimization methods, dedicated to solving such problems, is one of the most voluminous and complex. But these difficulties are technical, and in principle the task is simple: you need to maximize the criterion under the given constraints.

The task is significantly complicated when moving from a single to several criteria. It would even be more correct to say that a multicriteria task is fundamentally different from a single-criterion task. This is manifested in the example of the two-criterion task (see Figure 5.21).

In comparing alternatives x_1 and x_2, or x_1 and x_3, then there is no doubt since both, x_2 and x_3, by both, q_1 and q_2, criteria are better than x_1. But how to choose between x_2 and x_3? Each of them is better than the other by one criterion and worse by another.

In the theory of choice there is a story like an anecdote, in which the mathematician was asked to describe the algorithm for getting tea. "It's simple", he replied, "It is necessary to pour water into the kettle, put it on the fire, bring it to a boil, and throw tea leaves into it. In three minutes the tea is ready". And if they give you a kettle with water? "It is necessary to pour water from the kettle, and the task comes down to the previous one", was the answer. The story is that attempts were made to solve a multicriteria task by reducing it to a single criterion (or a sequence of single-criterion tasks) since the way to solve the latter is obvious. Several methods have been developed, of which the following has (unfortunately) become common.

Building a "super criterion", a "global criterion" $q_0(x)$ as a combination of local criteria $q_1(x), \ldots, q_k(x)$:

$$q_0(x) = f\left[q_1(x), q_2(x), \ldots, q_k(x)\right].$$

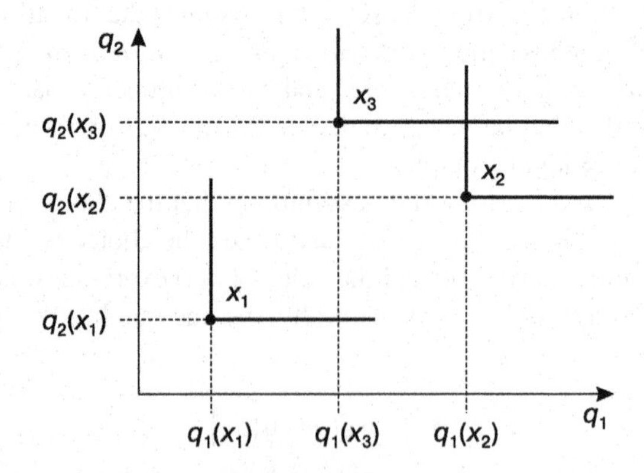

FIGURE 5.21 An example of a two-criteria choice.

Along with the technical difficulties of combining criteria measured in different scales (the complexity is solved by their artificial reduction to one scale), everything depends on the choice of the ordering function *f*: its task will lead to the choice of a single alternative, but the choice will be different when moving to another ordering function. One feels the presence of undesirable, but inevitable arbitrariness.

Conditional optimization, in which one, "most important" criterion stands out, the rest are transferred to the category of conditions, that is, fixed at an acceptable level for the customer. A variant of this task is to set conditions in the form of inequalities. In this case, there is an arbitrariness of the resulting decision, which depends on the specified conditions.

The method of concession, in which the criteria are ordered by importance, and then optimization is performed on the most important criterion. After this, a concession is assigned for this criterion, that is, the value by which we agree to lower the achieved value of the first criterion to maximize the value of the second within the limits of this assignment, and so on. Here arbitrariness is present in the form of ordering the criteria and values of concessions for each of them.

Lexicographic ordering. Unlike the method of assignment, the criteria are considered very different in importance that the application of the following criterion is made only if the previous one gave an ambiguous answer, and without any concessions. The term "lexicographic" is used in connection with the fact that this principle is used in dictionaries: there the ordering of words corresponds to the order of the letters of the alphabet in the word being searched for.

The method of setting the level of claims. In contrast to the previous methods, in this case, it is not a search for the best (in one sense or another) alternative but the prescription of its desirable qualities and a check whether there are any available alternatives *X*. If the answer is positive, it is desirable to specify the existing superior alternatives, if the answer is negative, indicate the closest to the specified criteria.

Although the International Symposiums of the Multi-Criteria Decision Making (MCDM) Making) are held every two years, where new versions of the above methods are discussed, we note that all these methods are attempts to apply single-criteria thinking to a multicriteria case. Inertness of thinking forces one to search for the single right decision, whereas in the multicriteria case such, as a rule, does not exist.

Meanwhile, an adequate solution to multicriteria problems was proposed as early as the beginning of the last century by the economist Pareto. It is based on the fact that *the preference of one alternative over the other should be given only if the first by all criteria is better than the second*. If the preference at least by one criterion is at odds with the preference by another, then such alternatives are considered *incomparable* and equally preferable.

In Figure 5.21, we introduce the notions of dominant and dominated alternatives. An alternative that by all criteria is inferior to the others ($x_1 \prec x_2$; $x_1 \prec x_3$), it is said to be *dominated*, and superior to all criteria, *dominant*. Now the choice in a multicriteria case becomes obvious: all dominated alternatives should be discarded. But the result is generally ambiguous, for example, in the case presented in Figure 5.21, the selection results are x_2 and x_3; there are no better options for both criteria, but they are incomparable.

The set of nondominated alternatives is called the *Pareto set*. This is an adequate solution to a multicriteria task.

However, in real life, only one option can be realized, and the question arises: which of the options from the Pareto set should be implemented? There is a question of choosing on the Pareto set. Its elements are incomparable, that is, they are the same in the sense that they are better for all the criteria, so you can choose any. There are different ways to choose in this situation.

A strong-willed choice: the decision-maker independently determines which option to exercise, or uses the services of experts.

Random choice: the decision is given to the will of the case (throwing a coin, a dice, etc.).

Introduction of additional criteria distinguishing alternatives from the Pareto set (in particular, application of a global criterion or introduction of a new one).

An interesting option of choice on Pareto set was suggested by D.S. Hammond, R.L. Keeney, and G. Raiffa [9]. They proposed to overcome the difficulty of comparing and choosing between the data expressed in different measuring scales by the method of "equal exchange". The idea is not to compare the data itself (they are incomparable in principle) but to compare the gain or loss that results from the preference of one to the other.

The method was developed for choosing among several alternatives by a combination of several criteria, a situation typical in management practice. The method can be presented as the following algorithm:

1. the situation is represented by a two-dimensional "Table of consequences" (an analog of the table "Object–Properties"), which vertically lists the criteria ("goals") and horizontally compared options. The table is filled in as a whole; the case of missing data (unfilled table cells) is not considered;
2. all dominated alternatives are eliminated;
3. if the values of some criteria for all nondominated alternatives coincide, this criterion can be excluded from consideration as irrelevant for selection;
4. an attempt is made to bring some unequal criterion to the same value for all alternatives so that it can be excluded from consideration. It is proposed to do this using the "equivalent exchange" method, increasing the value of one criterion, and reducing the value of another by an equivalent (for losses) value. If this succeeds, this criterion is excluded;
5. Stage 4 is applied to each criterion until there is one alternative or one criterion by which the choice is made.

The main difficulty of this technique is to determine the cost of changing the value of each criterion. For example, when choosing between flights of different airlines, it is difficult to exchange safety degrees for convenience of departure time. In such cases, it is recommended to consider more easily comparable pairs of criteria. It may happen that by making the simplest exchanges, you get a solution, and you do not need to wrestle with complex exchanges.

The main result of this section is that for a multicriteria problem, there is no unique correct solution; there is some (Pareto) set of acceptable solutions from which one can choose any.

Choice based on pairwise comparisons. In real life, there are often cases when no criteria allow to select the "best" alternative. For example, a boxer can measure weight, muscle volume, determine the reaction rate, etc., but it is impossible to predict from this data whether he will become a champion. In such cases, the criterial language loses its meaning, and with it the corresponding methods become inapplicable.

However, although an adequate assessment of a separate alternative in this case is not possible, it is possible to put two alternatives in such a competitive situation, where in reality they would compare their qualities, and the outcome of such a competition will determine their order of preference. Examples of such situations are tournaments, contests, fights, battles, or any pairwise comparisons.

If there are more than two alternatives, then the question arises: how to distinguish among them the most preferable if we only have the results of pairwise comparisons? With respect to such a task, quite extensive mathematical theories have been created since many alternatives can be finite, countable, or continuous, and the relationships between pairs themselves can be described differently. For our purposes, we confine ourselves to the representation of pairwise comparisons by the so-called preference graph.

The preference graph is a drawing that is obtained as follows (see Figure 5.22). In Figure 5.22:

1. circles represent alternatives;
2. they are numbered (these will be the vertices of the graph);
3. if any two alternatives are compared, a line is drawn between them (called an edge or an arc of a graph);
4. if by comparison one alternative "won", this is indicated by an arrow in the direction of the loser;
5. if the outcome is a draw, the line remains undirected.

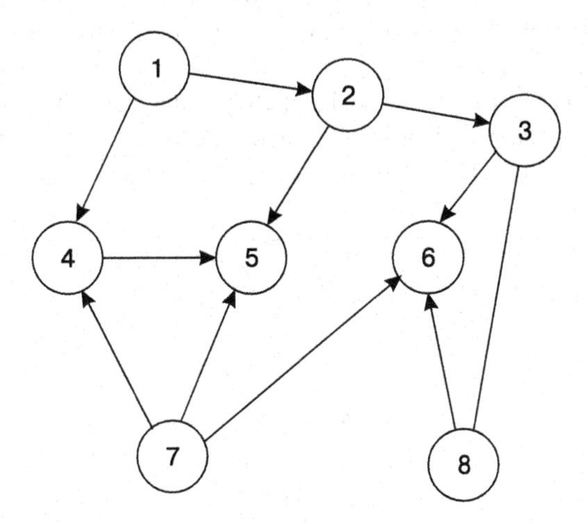

FIGURE 5.22 Graph of preferences between alternatives.

With such an observation protocol, one can single out the "best" alternatives. For this, it is necessary to determine the criterion of who is considered the "best", which can be done in different ways. For example, consider the best one who has not lost even once. Then alternatives one, seven, and eight will stand out. You can (to distinguish between them) take as a criterion the number of "fights" won; then the seventh alternative will be the best. But objection may be raised on conducting varying number of battles for different participants. It becomes clear that for a fair comparison, it is necessary to hold meetings of "everyone with everyone". However, it cannot be "the best" by the chosen criterion (e.g., there will be no one who has not lost once). It is necessary to enter other criteria. But the main obstacle for obtaining a complete set of pairwise comparisons is their large number, $N(N-1)$ with large N; therefore, it would be impossible to determine the world champion in any sport. True, the athletes have developed shortened, approximate methods for determining the leader: either zonal competitions with subsequent battles between the winners of the zones, or the Olympic system with a knockout after the first defeat.

On the general theory of choice. In the real practice of choice, there are cases when the main assumption of the theory of pairwise comparisons is not fulfilled. It consists in the fact that the order of preference in a pair is determined only by the qualities of the compared alternatives and does not depend on the presence or absence of other alternatives. If this is not the case (e.g., the choice between ground coffee or beans depends on whether you have a coffee grinder), then the language of pairwise comparisons loses its meaning. It makes little sense to build comparison theories based on the relationship between three, four, etc. options.

A very high degree of abstraction of the language of "global functions of sets" is proposed based on the concept of the choice function. This function $C(X) \subseteq X$ has as its "argument" the whole set X of alternatives $x \in X$, and its "value" is a certain subset of the set X: from the empty set — rejection of the proposed, to the whole X — "take all" (see Figure 5.23). By presenting certain requirements to $C(X)$ functions, various situations of choice can be described (including those discussed above). The selection function turned out to be a little studied and very complicated mathematical object; we will not go into details and limit ourselves to the abovementioned only for completeness.

Collective choice. Of the many tasks of choice, the task of multilateral decision-making is of particular practical interest, when a choice is made not by one person, but by a group of people. This assumes the highest degree of agreement between the

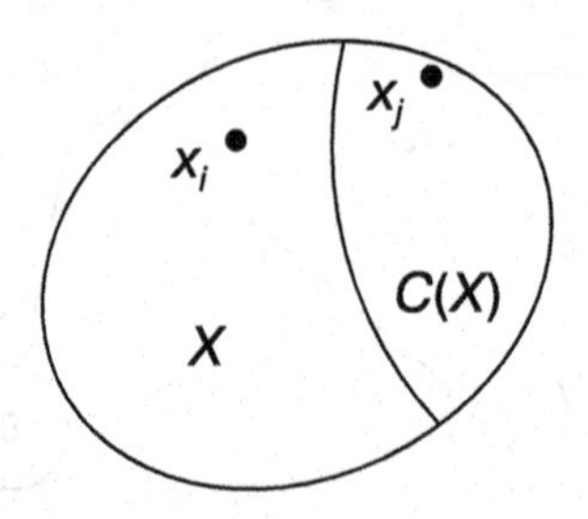

FIGURE 5.23 The choice function for acceptable alternatives.

members of the group regarding the common goal, and the choice has to be made between the options of means to achieve this goal.

A typical example is the election to a leadership position. Of the several candidates for this post, only one can be elected, and each voter is free to express his/her personal preference. The group decision $C_0(X)$ about the most preferable candidate $x_j \in X = \{x_1, \ldots, x_k\}$ is obtained by "recalculating" all individual preferences $C_1(X)$, $C_2(X)$ …, $C_N(X)$, where N is the voters number, into one "collective" preference $C_0(X)$ with the help of the procedure f: $C_0(X) = f\,[C_1(X), C_2(X),\ldots, C_N(X)]$, announced in advance and adopted by all members of the group. Such an operation is called *a voting procedure*. It is of great interest to find out what properties voting procedures have, both because of their practical significance and their diversity (the function f can be set differently). Moreover, often the results of voting are unexpected and sometimes undesirable. At the same time, many believe that by changing one voting rule to another, you can avoid the "wrong" result not only now but also in the future. It's a delusion. It is necessary to understand the nature of voting well to use it properly.

Seven paradoxes of voting. "One mind is good, but two is better", says the saying, suggesting a case in which both minds (and by induction, a larger number of minds) with the same intentions try to find a good choice.

When there are differences of opinions in a group, voting is the only possible way of forming a "common" opinion, that is, a collective decision. But voting procedures have a number of properties, in some cases giving unexpected or undesirable results. We list these properties and call them *paradoxes of voting*.

1. *The group is not always right*. The team consists of subjects, each of which may be mistaken. This leads to the fact that voting, even unanimous, does not guarantee the correctness of the decision. It is necessary to note, nevertheless, that due to the mutual compensation of opposite opinions, the probability of an error of collective opinion is less than the "average individual" one, but it remains nonzero. There are cases (Bruno, Galileo, Copernicus, and others) when one dissenter possessed the truth, while all the others were mistaken. Hence, voting is intended not to get truth, but to coordinate the actions of the group after the vote: all members of the group obey the decision, even if someone do not agree with it.
2. *The possibility of failure to make a decision*. Although voting is meant to make a decision, any voting procedure may result in that the agreed terms of decision-making are not met and, therefore, no decision will be made.

   Let's explain this with examples. Say, "a simple majority" (50% plus one vote) will not work if the votes of an even number of voters split equally. The amendment "to the chairman — decisive vote" bypasses this situation, but if an odd number of voters shares in such a way that the chairman is in half, less than one vote, the question arises: what is the "decisive vote"? When adopted by the "qualified" majority (two-third) on the scientific councils, there were cases that the defender lacked a small fraction of the vote. Even with the principle of unanimity (consensus, the right of veto), a decision may not be made. This is a feature of all voting procedures.

3. *The Condorcet Paradox* (so named after the mathematician who explained this paradox). The essence of the paradox is that the preference graph is cyclical (see Figure 5.24). For example, let each of the three factions of parliament, which form the majority only in pairs, propose their own version of the bill: *a*, *b*, and *c*. Or three guys argue whose girlfriend is better and intend to resolve this issue by a voting. If this trinity has its individual preferences as follows: $(a \succ b \succ c)$, $(b \succ c \succ a)$, and $(c \succ a \succ b)$, then they fall into the Condorcet paradox. Any procedure will either end with a rejection of the decision (since with such preferences there is no solution), or with an artificial interruption of the procedure (e.g., in a pair comparison according to the Olympic system), the outcome will depend on the sequence in which the pairs are considered.

   Sometimes the Condorcet paradox is insignificant (if the cycle of preference turns out to be at the bottom of the chain of many alternatives and does not affect the choice of leader). If it is necessary to resolve it, the way out may be in convincing one of the voters (now it is called black or white PR, "public relations") to change his ordering alternatives, without changing superiority of the own candidate. The cycle in the graph will disappear, and the solution will be the only one.
4. *The possibility of the victory of the minority under the majority voting system.* Let the decision be made by a majority vote (this is the majority system). It turns out that there are opportunities for the legitimate victory of a minority, and there are several such opportunities.

   The first is the recognition of elections legitimate (legal) with a low (less than 50%) electorate attendance. The solution is automatically granted to the minority. It is difficult to condemn such a situation since nonparticipation in elections implies indifference to what decision will be taken; then let it be made by those to whom it is not indifferent.

   But the minority can win even with 100% electorate turnout.

   The second such opportunity is the "pilfering" of votes. Let's explain this by example. Let one coalition possess 60% of potential votes, the second belongs to 40% of the electorate (see Figure 5.25). If the first nominated two candidates, and even equivalent, and the second nominated one, the minority will win. The reasons for pilfering of votes may be different, but the result is the same.

   However, the minority has a chance to win at 100% turnout and without stripping of votes. Let us explain it again with an example. Let the decision be made by a majority of two-third. If in the end a minority representative

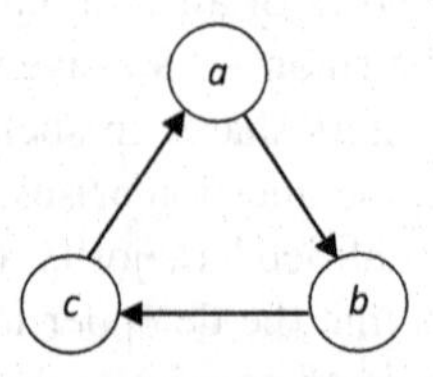

FIGURE 5.24 The cause for Condorcet paradox.

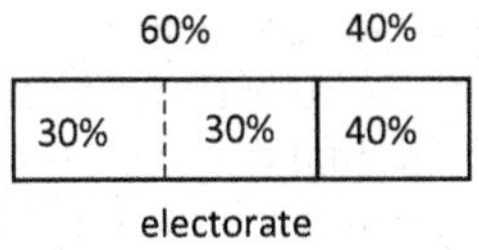

60% 40%

30%	30%	40%

electorate

FIGURE 5.25 An example of pilfering votes.

won, then at the last stage of the procedure he gained a majority. If the participants of the last voting stage themselves were chosen according to the same rule, then the situation depicted in Figure 5.26 arises. It is easy to see that the minority won in 4/9 against 5/9. For the realization of this possibility, it is necessary to fulfill three conditions:

a. elections must be multistage (since at each stage the decision is made by majority vote);
b. the minority must observe the discipline of voting (i.e., to vote exactly where the organization of the whole case requires: if at least one of them changes places with the enemy, according to the dotted arrows in the figure, then nothing will come of it);
c. the minority should be large enough to ensure its majority at the last stage. If there were not four, but three representatives of the minority in the bottom row, again nothing would have happened. However, minority interest may be less if additional voting levels are introduced. So, if we supplement the scheme of Figure 5.26 on another level, the proportion of minority in 4/9 (44.4%) will decrease to 8/27 (33.7%). This scheme is not only of theoretical interest: multistage voting schemes are used in life, for example, the two-step procedure defined by the constitution for electing the President of the United States has already five times out of 45 resulted in the victory of minority candidates. For instance, it was in 2002 in the rivalry between Bush and Gore: the first won, with only 48% of the votes in the primary elections.

In another form, the paradox of the majority system is laid down in the electoral law of Australia. An interesting difference is that there is a one-time vote (i.e., physically one step). However, unlike us, the Australian voter is obliged not only to indicate which candidate he gives primacy to but also to enumerate all the other candidates in order of preference.

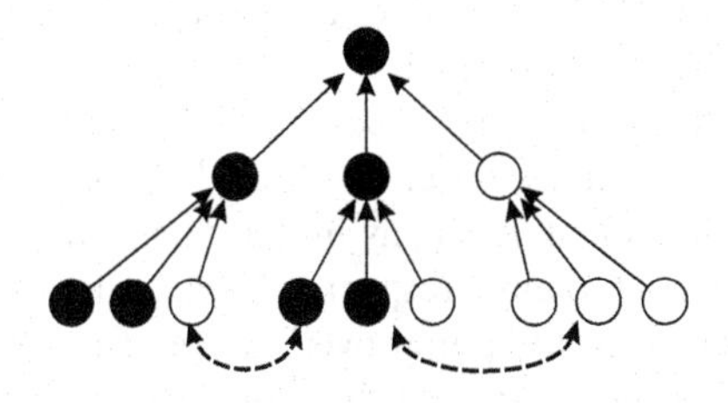

FIGURE 5.26 An example of minority winning in majoritarian voting.

Bulletins are processed according to the following procedure (which includes the paradoxes already known to us). Of all the ballots are extracted candidates who received the highest score (No. 1), as well as gained the votes necessary for the election threshold percentage of votes passes to parliament. In the event that none of them got such a percentage (the option of pilfering votes!), the counting commission takes a sample of all candidates who received a second number on the ballot and repeats the check to exceed the set threshold and so on to the level at which someone reaches the threshold (as we see, instead of physical multistage, algorithmic is implemented). In the Australian Parliament, there are parties that have received a very low percentage of votes. Their chance is that the votes will be pilfered at all but the last level. The discipline of minority voting is that they follow the instructions of their leadership to put their candidate on the ballot in the last place. If the procedure goes down to the last level, we are all there, plus all those who put us in last place. This is quite enough to overcome the barrier to obtaining a mandate.

5. *The paradox of the overwhelming majority.* Many believe that when voting on the principle of "one person one vote", the greater the percentage of votes gains an alternative, the more democratic is the decision. It's a delusion. Apparently, this impression is based on the fact that politicians feel all the more confident when most of the electorate supports them; the more they feel that they are representatives of the people.

   The paradox is that such an impression is psychologically understandable since it is based on the common notions of "ours and others", "us and them", but it has nothing to do with the concept of democracy. Whatever high percentage of the majority is appointed for the legitimacy of the decision, the decision is not democratic. Let us explain this with a simple example.

   We will offer the most "democratic" voting procedure, consisting of only two rules:

   a. the decision is made with any number of N voters only if at least N –1 people voted for and only one (no more!) "contra". (We emphasize again: N can be arbitrarily large)
   b. everyone votes "pro" if the proposed alternative does not personally harm him (and even more so if it is beneficial to him).

It seems impossible to propose a more "democratic" procedure. But if society approved it for collective decision-making, it said goodbye to democracy. Now the chairman can (if he wants) through this procedure implement any decision he personally pleases.

For example, let us use this procedure to decide whether we all should go from one state to another. Let the "state" denote the presence of each certain amount. Statement: from any initial state, using the rule introduced, you can be transferred to any prespecified state in a finite number of steps. For clarity, let me want to "pump over" all your money into one pocket. Step one: who is to take away all the money from certain person (name) and distribute it equally to all? The outcome is clear. It is possible to speed up the process and to suggest that money be taken away from everyone in turn and given to the target person. The procedure *here* will work. Sooner or later the goal will be achieved, and will be quite legitimate.

Do not think that this example is artificial. Worse, in the practice of applying the decision by the "overwhelming majority", it was accompanied by the elimination of the disaffected. Suffice it to recall the actions of dispossession in the 30s of the last century in Soviet Union, decisions about which were taken by the committees of the poor. Or a story of a priest in Germany (at the same years) under Nazis: "When they start to exterminate Jews, I was silent: I'm not a Jew. When they wiped out Communists, I was silent: I'm not a Communist. But when they put me in concentration camp, there was nobody to plead for me".

The essence of the paradox is that this procedure legalizes sacrificing the interests of one to all others. In this case, the others forget that each of them can become the same victim.

Thus, majority voting and democracy are just different things. Voting is only a procedure of making a collective choice. The essence of democracy is not that everyone can participate in direct and secret elections. Decisions can be made both collectively and individually, and democracy is to ensure that at the stage of execution of a decision, the interests of any minority are protected, and above all, the basic rights of each individual are protected (right to life, right to property, right to freedom). However, the periodicity of the majority elections makes them of democratic nature: in the next elections, the people can correct their mistake if the policy approved by them earlier turned out to be flawed. Incompatible ideals of democracy — liberty and equality — are reconciled by periodic change in power one of two political parties holding opposite ideals.

6. *Paradoxes of unanimity.* If democracy is defined as protecting the interests of everyone, then the only democratic voting procedure is unanimous decision-making: at the decision-making stage, everyone can defend own interests by voting against an alternative that is not suitable for him or her.

There are important practical situations in which the principle of unanimity applies: the right of veto in some parliaments; decision-making by the UN Security Council; election by cardinals of the next pope; the verdict of the guilt of the defendant by the jury; decision-making in joint stock companies with unlimited liability. The same principle is strongly recommended to be followed in applied systems analysis since its ultimate goal is the creation of an improving intervention.

However, in this case, paradoxical situations arise. First, sometimes the principle of unanimity ("all for") is replaced by the principle of consensus ("no one is against"), whereas these are different things: abstentions are identified with consonants, the absent are excluded from being taken into account. A vivid example is the decision of the Security Council to conduct a war in Korea under the UN flag taken in the absence of a representative of the USSR.

The second paradoxical situation arises when the desired solution cannot gain 100% of the votes. There are at least two ways to try to reach agreement in such a situation.

The first is to find a compromise. Let us illustrate this with a diagram in Figure 5.27. If we draw in circles the set of acceptable alternatives of the three decision-makers, the failure to achieve unanimity is simply due to the lack of an alternative, which is acceptable to all. The solution may be that someone (voluntarily or under the influence) expands his circle of acceptability so that mutually acceptable options appear (the dotted line covers them).

The second method can be called the "staircase method". If we can't jump on the barn, we set up a ladder and climb up the steps. So here, you can try to approach the desired, but not immediately achieved unanimously goal, by small steps, each of which is implemented unanimously. An interesting example of such a real attempt is presented by R. Ackoff from his practice [4]:

> Consensus is often difficult to achieve, but rarely impossible. I found that in difficult cases the following procedure is very effective. The first is to clarify as much as possible the wording of alternatives between which consensus does not reach a choice. The second is to collectively build an alternative effectiveness test and decide by consensus that this test is fair and that everyone agrees to follow its result. The third is to conduct a test and use its result. I was able to successfully apply this procedure even in such a case when the legislators of one state could not agree on whether to impose or not the death penalty for murder. As a result of the discussion, members of the legislature agreed that everyone has the same goal — to minimize the number of murder victims. As soon as such an agreement was reached, the problem was reduced to a specific question: does the introduction of the death penalty reduce the number of murders? All agreed that it was necessary to conduct a study that answers this question. Such a study was carried out, and its results were used (it showed that the number of murders in a number of states did not change noticeably and significantly before and after the abolition or introduction of the death penalty).

If it is not possible to reach agreement not only on the alternatives themselves but also on the way they are tested, then, according to R. Ackoff [2], a consensus decision should be found on what to do next. It is interesting to observe

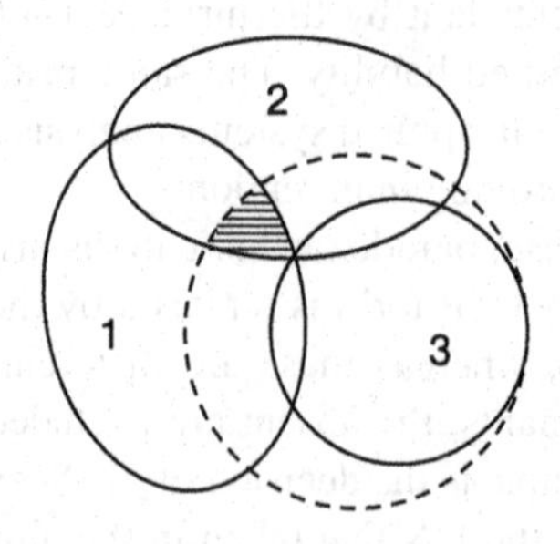

FIGURE 5.27 Scheme of finding a compromise.

that in such cases it was usually decided to entrust the choice to one of the authoritative and responsible persons (we will come back to this point later).

In general, the paradoxes of unanimity are associated with a delusion about its nature. There is an opinion that unanimity is a sign of rightness of the decision: the closer we are to unanimity, the closer is the truth. This misconception was discussed when considering the first paradox. Such an opinion is based on a real increase in the *statistical* reliability of the solution with an increase in the sample size, which, however, does not guarantee its *absolute* correctness.

This does not mean that the desire for unanimity and consensus in principle is wrong and should be rejected. It plays a significant role in our world. But, if one does not take into account that it inhibits serious disputes and objections, this may lead to undesirable consequences, such as collective pathologies of "clan thinking" and "group thinking" (see the twelfth stage). It restricts, rather than expands, the diversity of opinion and narrows the scope for discussion. It is necessary, but not sufficient, for life in our complex and changing world. Other ways of thinking and acting are needed. After all, it is known that the more important and complex the question, the harder (or impossible at all) to achieve common agreement.

7. *Arrow's theorem on impossibility*. The most general theoretical results on collective choice were obtained by Arrow, for which he was awarded the Nobel Prize in Economics. The most famous was his "Impossibility Theorem". The question in it is this: is it possible to say something meaningful about all elections procedures? The answer is: not all of them, but about the "good", "acceptable" procedures, satisfying certain reasonable requirements, one can try.

   From the various functions F of individual choices, $C_0(X) = F\,[C_1(X), \ldots, C_N(X)]$, we select those that meet the requirements expressing our understanding of what is "correct", "fair", and "good" collective choice. In fact, there are only four such requirements (there are others in the original, but they are purely technical and mathematical, about amounts of voters and alternatives):

   a. all individual preferences of $C_i(X)$ must somehow be taken into account; there should not be such an individual whose opinion is made mandatory for everyone, regardless of the opinions of others (the function $C_0(X) \equiv C_1(X)$ is called "dictatorial", and this condition expresses the undesirability of dictatorship);
   b. if, as a result of a group choice, an alternative was given preference, then this decision should not be changed if someone who had previously rejected it changed his opinion in its favor (the condition of monotony);
   c. if changes in individual preferences have not affected any alternatives, then in the new group ordering the order of these alternatives should not change (the condition of independence of alternatives). Let us explain this requirement by an example. Let $C_0(X) = F\,[C_1(X), \ldots, C_N(X)]$. Let us mentally go back and remove the ballot of the i-th voter

from the ballot box and ask him to "think it over again". Let him in his preference swap two candidates. We will recalculate $C_0(X)$ taking into account another variant of its bulletin. The result can often remain the same. But if for these two candidates the situation was unstable and one vote turned out to be enough to change it, then it would be fair that in the new ordering the change would affect only these candidates and not hurt the rest;

d. for any pair of alternatives, such two sets of individual preferences are possible for which the order of these alternatives is opposite ("sovereignty condition").

This is the part of "If ..." in the Arrow theorem. The part "Then ..." reads (because of what it received the name "Theorem on impossibility"): the specified requirements are incompatible, that is, there are no voting procedures that meet all these requirements.

This came as a big surprise (after all, the requirements seem so natural and necessary!) and caused heated debate. It turned out that the reason for such a result is the paradoxes mentioned above, and in the first place came the fact that the collective choice could "get stuck", end up with no decision-making; and the sole, "dictatorial" choice never. This led to a lot of noise around the impossibility theorem: "Science proves the weakness of democracy," "Science proves the inevitability of dictatorship", etc. Now the dust has settled (many years have passed) and comments on the impossibility theorem can be made like this:

a. whether someone likes it or not, this is the nature of the vote ("I didn't like it when I fell and hurt myself, but that would not disaffirm the law of gravity");
b. the Arrow theorem is about voting, not about democracy. These are different things, and its political interpretation is a substitute for concepts;
c. failure to make a decision will lead to losses, and losses may be acceptable or intolerable;
d. if losses are acceptable, we prefer to make decisions collectively by voting: this gives a certain meaning to our social activity;
e. if losses from the failure to make a decision are intolerable, the very possibility of failure to take a decision should be excluded. This can be done in only one way — to move to the sole decision-making, that is, to the dictatorial function;
f. in itself, the sole decision-making is neither bad nor good. It all depends on the specific conditions. For example, to abolish the principle of unity of command in the army means to weaken the combat capability of the army (which is proved by the experience of the Red Army, until Zhukov did not get Stalin to abolish the diarchy of commanders and commissars during World War II). Yes, and in everyday life in a collectively uncertain situation, we resort to the opinion of authoritative, trustworthy, prestigious members of society;
g. discussing the properties of voting procedures has nothing to do with politics. This is only a strict logical consideration of the peculiarities

of the recalculation formulas of individual preferences into one, called collective. How to use the knowledge of these properties in real life is a matter of policy.

Decision-making in a social system. So far, we have considered decision-making *analytically*: there are many alternatives, from which it is necessary to single out one, the most preferable. The focus here was on internal, technical side of the matter: how to make such a choice. Many options were found, and in each situation, the decision-making algorithm had features that took into account the specifics of the situation.

Now we turn to the *synthetic* consideration of choice, that is, consider what are the external conditions for decision-making (see Section 3.2 in Chapter 3).

Decision-making is the most important function in government and management. Governance is the implementation of change in a controlled organization. An organization is a certain structure with the distribution of rights and duties, power and responsibility. Therefore, decision-making in such a system should occur taking into account the authority of the decision-maker. These powers are determined by his position in the organizational structure. At the same time, it is necessary to take into account the position of the decision-maker, both in the structure of his organization and in the structures of external systems, to which our organization belongs.

Let's start by considering the specifics of making internal decisions by a person who is a member of an organization. The range of his decisions is to a certain extent limited by his job description. But what decisions, and at what level of hierarchy should they be made? The desired goal is to minimize the likelihood of making erroneous decisions. Exposure to error is an inevitable feature of any subject, and at any level of the management hierarchy. Hence, the conclusion that the decision should be made at the level where the maximum information is concentrated on the situation requiring intervention. This is not necessarily the top management in the organization, which is the basis for the need to delegate authority to lower levels of the hierarchy.

This redistribution of power and responsibility may take many different forms. For example, a subordinate is given the right to veto decisions of higher (the seller has more information about buyers than the store director). Another option is implemented in the airline SAS. All employees, regardless of rank, are required, upon receipt of a request, offer or complaint from a client, either to respond accordingly or to see that this is done properly by the relevant service. Shifting responsibility to others is prohibited.

So, there is the task of transferring the right to make decisions to those lower levels that are most competent in the problem situation. The head is usually reluctant to give a part of his authority to subordinates. There are several ways to convince him to do this. The main thing is to show that the organization will function better if you give employees more freedom in decision-making. Employees in large organizations are usually alienated because they do not have the right to vote, they do not affect decision-making, and are largely similar to robots.

Another important feature of the organization's internal decision-making environment is the education of employees. The more educated the workers, the less efficient powerful management of them. A good example is universities, a significant proportion of whose employees have advanced degrees. No rector runs a university, deciding what should be taught and how, how to be assessed, which books should be used. All important decisions for the educational process are made at the lower levels of the hierarchy.

Yet another way to encourage senior managers to delegate authority to subordinates is to familiarize them with the problems of the lowest level, users. Leaders often believe that their power is manifested in looking all-knowing and sublime; that communication with inferior ones reduces their authority and status; they often don't want to know what is happening at the bottom. Useful advice is that the head of the transport company sometimes travels along with ordinary passengers, the store director himself makes purchases in it, etc. For example, the most long-term rector of Tomsk University, Prof. A. P. Bychkov, personally attended the lectures of all professors in all faculties of the university, at least once.

Summing up the topic of internal decision-making in an organization, one can say that the modern tendency of management development goes in the direction of decentralization of management, in the transition from administrative commanding leadership style to participative, participatory, and democratic management. Although the dictatorial management style is not excluded (e.g., in conditions of war, emergency situations, etc.).

Let's now turn to the problems of decision-making related to the need for interaction with external management structures. If internal problems are connected with the optimization of the distribution of rights and duties between subordinates, which is in the power of the manager, then relations with external structures most often boil down to asking for decisions of higher-level managers on issues that are essential for the subordinated organization.

There are two types of interactions between social systems: either they give us the opportunity to do or get something that we cannot have without their assistance, or they prevent this possibility. In other words, the systems with which we interact expand or reduce the number of our allowable actions: they either allow or forbid to do something. Although there are systems whose operation is predominantly restrictive (e.g., prison) or predominantly broadening (e.g., library), many combine these functions (e.g., schools, government agencies). If some system is focused on bans, it is called bureaucratic. R. Ackoff defined the bureaucrat as having the right to say "no", but who could not say "yes". There are two reasons for this. The first is that in a bureaucracy "no" does not lead to what is considered an error, only "yes" can do it. Therefore, the best strategy, in order not to make obvious mistakes, is to reject the proposals, not allowing anything to be done. The bureaucrat cannot say "yes, since to do this he must forward the request to his superior so as not to risk his disapproval. But it means to recognize the limitations of their importance.

The second peculiarity of the bureaucracy is the establishment of unified indestructible rules and restrictions that do not allow exceptions. Thus, bureaucrats are often busy with not allowing others to do something. They say exactly what to do, depriving the performer of choice and initiative. Violation of instructions is a grave sin; attention and sympathy for the petitioner are alien to the bureaucrat (as it increases the risk of error and subsequent punishment) [4].

Conclusions. We considered only a few problems of the theory of choice. The selection criterion was a frequent occurrence in practice and the upcoming use of the results in presenting the problem-solving technology. The reader should know that they constitute only an insignificant aspect of all variants of the decision-making practice, a partial list of which was generated by the morphological analysis conducted at the beginning of the description of this stage. For example, when faced with the need to choose in conditions of uncertainty, one should identify its type and turn either to game theory (with uncertainty of ignorance), or to the theory of statistical solutions (with stochastic uncertainty), or to the theory of fuzzy sets (with vague uncertainty). In the absence of uncertainty of the consequences of the choice made, the problems are solved by optimization methods. For each of these options, there is an extensive scientific and educational literature. A brief overview of the ideas contained in them and a bibliography is in the book of F.I. Peregudov and F.P. Tarasenko [5].

Questions and Tasks

1. Define "choice."
2. Why is it impossible to create a universal theory of choice?
3. How to solve multicriteria problems correctly?
4. What is the "Pareto set"?
5. What are the difficulties of choice based on pairwise comparisons?
6. List the seven voting paradoxes.
7. What are the features of decision-making with respect to the surrounding social environment?

5.2.12 Stage Twelve. Implementation of Improving Intervention

Abstract

Planning of ways to bridge the gap between the purposes and the problem messes was carried out in stage 10 ("Generation of Alternatives"). The choice of intervention to be implemented was made at stage 11 ("Decision-Making"). Now it is time to plan the resources required to implement the designed intervention. During this planning phase, each type of resource must be defined.

The next phase in the implementation of the intervention is the organization of the execution of the decision. This action is essentially an act of governance. In accordance with this situation, one of the seven types of controls described in Chapter 4 is used, or a suitable combination of them in the desired sequence.

After the decision about what kind of improvement interventions should be implemented (this is the result of the previous stage), work is needed to implement this decision (this is the objective of this stage). But between the decision-making and its implementation, as they say, "the distance is huge." This distance is overcome by planning the necessary actions and their execution when monitoring the course of events and making amendments where necessary.

The planning of the final results was carried out at stage 6 ("Targeting"). Planning of ways to bridge the gap between the purposes and the problems messes was carried out in stage 10 ("Generation of Alternatives"). The choice of intervention to be implemented was made at stage 11 ("Decision-Making"). Now is the time to plan the resources required to implement the designed intervention. During this planning phase, each type of resource must be defined:

1. how much of it will be needed, where and when;
2. how much of it will be available at the appointed place and time;
3. what should be done in case of its shortage or excess.

All types of resources should be considered:

1. people with their competencies and qualifications;
2. buildings and equipment (capital expenditures);
3. materials by types and volumes of reserves, and energy (consumable things);
4. money;
5. information;
6. time.

The next phase in the implementation of the intervention is the organization of the execution of the decision. This action is essentially an act of governance. In accordance with this situation, one of the seven types of controls described in Chapter 4 is used, or a suitable combination of them in the desired sequence. Usually after the restructuring (fourth type) regulation (third type) is applied, or overcoming complexities (second type). In any case, you will need to monitor the current state of affairs and identify assumptions and risks (*control* function). In addition, a subsystem for detecting and correcting errors in expectations and assumptions (the *learning* function) will be required to ensure adaptability to upcoming changes both within and outside the system.

Determination of assumptions and risks. No matter how well the project is planned and prepared, real events do not always happen according to the plan. Many external factors can affect the progress of the project and are beyond our control. Therefore, it is necessary to include this in the list of our assumptions.

Step-by-step progress along the chain from problem statement to its solution occurs through verification of the assumptions embedded into each stage: the transition to the next stage is carried out only if the previous one is considered successfully completed.

One of the roles of the systems analyst is to identify such external factors and, if possible, to integrate into the project by either counteracting them or monitoring their impact. Therefore, it is necessary to assess the likelihood and significance of possible circumstances, thereby contributing to the assessment of the riskiness of the project. Some of them will be essential to the success of the project, while others will be unimportant. An algorithm for working with assumptions may be proposed (see Figure 5.28).

For clarity, we give examples of possible assumptions for solving a social problem:

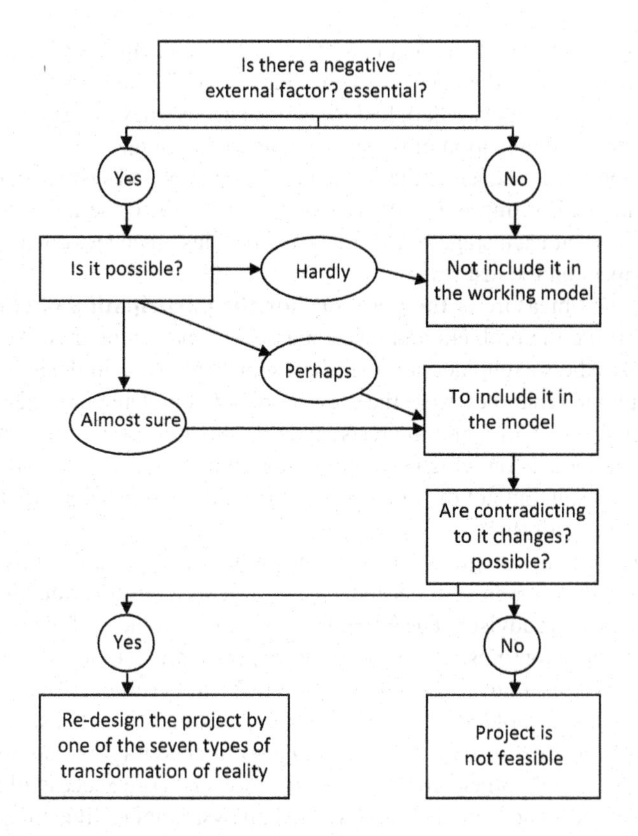

FIGURE 5.28 Algorithm of working with assumptions.

– local organizations will cooperate in planning our work;
– the necessary personnel are identified and hired, local and invited;
– sent to study the staff returned to participate in the project;
– necessary amendments have been made to the budget;
– the governing bodies have fulfilled the preconditions of the project sponsors.

In general, this stage is management.

Practice shows that the distance from decision-making to its implementation is not always successful. It is natural to try to collect and summarize the experience of success and failure in achieving goals. These efforts have led to the formation of a whole discipline, a modern teaching about management.

The proximity of management to systems analysis is so great today that the author has been lecturing on management (after the course of applied systems analysis) at Tomsk State University, promoting the view of management as an application of systems analysis to the problems of managing people in the organization, and the manager as a "systems analyst with a permanent place of work". We will not describe (even in abbreviated form) all the practical discoveries, heuristic guesses, and theoretical constructions that make up the achievements of modern management. They are presented in numerous textbooks on this subject.

Instead, let us note that the technology of the systems analysis embedded measures that contribute to the success of the final stage. Though these measures are distributed across all stages of applied systems analysis, they are aimed at ensuring the successful implementation of the decision at the last stage.

The key point, the essence of the whole ideology of applied system analysis, is the pursuit of the ideal of improving intervention. This leads to the need to implement special measures at each stage of the analysis, the consequences of which will have a positive impact on the last stage.

The first such measure is **the necessity for the participation of stakeholders**. While identifying the problem and target messes, we have already talked about the need to involve the stakeholders (or their best representatives) in the analysis. At the time, we explained that they were the only source of complete and reliable information about their own views and interests, and that this information was necessary to build adequate models on which the improving intervention will be based. This, of course, is important, but not the only or even the main reason. There are two equally important reasons for this.

One of them is the integration into the procedure of applied systems analysis of *training of stakeholders* in systems analysis. A systems analyst is fundamentally different from an expert advisor. The latter, faced with the problem of the client, sees its task in collecting symptoms, diagnosing, and writing a prescription as a doctor to the patient. The systems analyst sees his task not only in learning the necessary information from the stakeholders (the analyst knows what questions to ask, and only the stakeholders know the answers to them) but also in encouraging and creating suitable conditions to help others cope with their problems more successfully than they could themselves without his help. A systems analyst is more like a teacher than a doctor. A special term "facilitator" for such a professional has entered the practice of applied systems analysis.

The teacher cannot learn for their pupils: the pupils must learn for themselves. The task of the teacher is to give students the opportunity to learn more and faster than they can without his help. At the same time, participation of stakeholders in the process of analysis implements the most effective way of learning: not "by ear" and not "watching", but by "doing it yourself". Thus, applied systems analysis is primarily an educational process (not only for stakeholders but also for the analyst).

Since education is necessary for development, and it is impossible to learn instead of others, it is impossible to develop others. The only way of development is self-development. It is possible to support and help the development of others, but only with their participation. This is the reason for the need to involve stakeholders in systems analysis of their own problem mess. The difficulties to achieve this is no reason not to do this.

Another important reason for this is the fact that the designed improving intervention will not be carried out by a facilitator (systems analyst), but by the stakeholders themselves: the necessary resources, powers, personnel, finances, etc. are at their disposal.

In the implementation of plans will inevitably appear a variety of difficulties, faced with which, some give up ("objective obstacles stronger than us"), while others are trying in every way possible to overcome or circumvent them. The decisive factor here is often the one whose decision (someone else's or his) must be fulfilled by a person: he is more persistent and active in the implementation of his own goals. ("Who wants to achieve the goal — is looking for tools and methods, who does not want — is looking for reasons" — a well-known aphorism among managers.) Hence, it is very important to ensure that those who will have to implement the project of improving intervention want to do so. It is necessary that they feel like authors or coauthors of the project, that is, participate in its development.

It is especially important to make the participants of the analysis and actual developers, authors of the project of intervention — *the first persons of the problem-containing and the problem-solving systems.* This is so important that if it is impossible to involve these leaders to work on the problem, the chances of ultimate success are greatly reduced (R. Ackoff [2] even believes that they fall to zero, and recommends to cancel the contract).

The next measure to increase the likelihood of success is **to ensure voluntary participation**. Even if it is possible to gather stakeholders (or their representatives) to participate in the development of a way to solve the problem (improving intervention), this does not guarantee their active and effective work. Many factors affect how fully a person strives to realize their potential in the search for solutions. R. Akoff [4] believes that the most significant factor is *voluntary* participation of the subject in joint efforts (if participation is not voluntary, it cannot be effective).

What is the best way to achieve voluntary participation of stakeholders? Reflecting on the experience of his rich practice of systems analyst as problem solver, R. Akoff notes that the more willing they are to participate in efforts to develop a better intervention, the more confident they are in meeting the three conditions:

1. their participation will really affect the results obtained;
2. participation will be an interesting and enjoyable affair;
3. the results will indeed be implemented.

Let us briefly explain what these conditions are based on and how they can be fulfilled.

How to create confidence that participation will mean something? This condition is most fully realized if each of the participants feels equal in their decision-making. For example, if decisions are made by a majority vote and one of the parties concerned has such a majority, the other parties are unlikely to participate voluntarily. Increased confidence in the real impact on the course of the discussion can guarantee that everyone's opinion will be taken into account. Such a guarantee ensures decision-making during the analysis *only unanimously.*

How to make participation a pleasure? There are several ways to do this: encouraging humor and a friendly atmosphere, introducing entertaining elements to the work, giving serious work the form of play, and most importantly, the urge to imagination and creativity, which is exciting and enjoyable. Typical examples are brainstorming and idealized design (described earlier), evoking a deep sense of coauthorship.

How to present the implementation as feasible? One of the main conditions for increasing the confidence of stakeholders in the fact that they do not work "on the shelf" is the participation of the first persons of the systems involved in the problem situation, especially the problem-containing and problem-solving systems. Sometimes achieving this is not difficult, but it often requires serious effort.

The likelihood that these persons will subsequently actively fight for the implementation of decisions is increased if certain circumstances accompany the work.

For example, R. Akoff notes that implementation is more likely if the development of the recommendations was paid for ("it seems that people do not really appreciate what goes for free"). Even in cases when a worthy client is not able to pay for the necessary research, the way out is to find sponsors (various funds, state and municipal authorities, donors, patrons, etc.).

Further, the interest of the first person's environment (both his or her bosses and subordinates and others whose opinion is important to him or her); in fact, their interest is of great importance for ensuring the participation of the first person. Here can help efforts to inform the public, the different PR technologies.

R. Akoff attaches particular importance to the relationship of trust between the manager and the systems analyst; he even believes that such trust is absolutely necessary for the implementation of the results.

The head of the organization involved in the system analysis needs confidence that he and his organization will benefit from the implementation of the results of the study. Verbal assurances about the development of improving intervention are not enough, you need to trust the systems analyst, similar to that experienced by friends. The highest degree of trust is friendship. We believe that a friend will act in our best interests, even if he may suffer in some way. Therefore, the advice of friends, no matter how bad it is, people follow rather than the advice of enemies, no matter how good it may be. Friendly relations with the analyst provide the manager with a sense of security.

There are several techniques that contribute to the creation of, if not friendly, then trusting atmosphere between them.

The contract for work includes a condition that any party has the right to stop work at any time without explanation. This gives the manager a guarantee that if during the work there is a fear that its continuation may harm his interests, he can prevent it. (R. Akoff notes that in his practice this point has never been used by anyone, but it had an important psychological significance. My attempts to introduce such a clause in the contract with Russian clients ended with the fact that manager insisted on the declared right only for himself.)

The system analyst undertakes to train employees of the investigated organization to ability to carry out the systems analysis independently in the future. This increases the viability of the organization and is positively perceived by the manager.

The systems analyst not only does not expect and does not require recognition of his merits in the achieved successes; on the contrary, he strongly emphasizes the merits of other participants (which, incidentally, contributes to the growth of his authority).

System analyst openly proclaims and firmly observes the requirements to his professional activity (guarantee of access to any necessary information — with the obligation of confidentiality; guarantee of access to the heads of organizations involved in the problem situation; guarantee of scientific and business integrity; compliance with professional moral and ethical standards).

The systems analyst should openly and sincerely show respect for the intellect of the manager (of course, we are not talking about sycophancy).

Another measure to increase the probability of success of the transformation of the organization through systems analysis is **to take into account the peculiar characteristics of human beings**. The main difference between social and sociotechnical systems from other natural and artificial systems is the entry of people into them as the most essential components. People possess many properties inherent only to the subjects, unlike other objects of the real world; first of all, the ability to properly appreciate the interactions with environment, and (alas!) sometimes to be mistaken. In solving the problems of organizations we have to deal with these features and take measures to overcome their negative impact on the design and implementation of improving interventions.

There are several such features, such that J. Warfield [3] even calls them *pathologies*. It is necessary to distinguish such features for different human entities, for an individual, for a group of several people, and for a team of employees of some organization, which in number exceeds what we usually call a "group". The pathology of the individual can influence the behavior of groups, and pathology groups can influence the behavior of the individual. All of them together or separately can affect the effectiveness of efforts to transform the organization, the success of solving problems.

It is proposed to proceed with the following pathologies while working on the organization's system transformation.

Individual peculiarities. *The instinct of self-preservation* (instinct for survival). Inherent in every living being, it affects the behavior of the individual in a group or organization in various forms. It seems that this instinct can be blocked if the individual believes that all the cases in which he participates can be adequately performed only by joint efforts. In the absence of such faith, the instinct of self-preservation manifests itself in any action of the individual. For example, a typical manifestation of this instinct by managers is the fear of risk in decision-making. For really difficult problems, there is no single right solution. Managers prefer situations with one right decision as it reduces the risk of making a mistake. In the course of affairs, it may turn out that not everything turns out the way you wanted. So, the decision was wrong. But many managers are unwilling and unable to admit it. They are looking for someone to blame around them. But it is very valuable to recognize their mistakes and change course in solving the problem. Mistakes are not failures, but a source of learning and development.

Limit of attention (processing limit, "magic number"). It is established that a person can keep in the field of his attention only a small number of ideas or objects at a time. Different authors define this magic number differently: "seven plus-minus two", "five rather than seven", and estimate even "only three". The main difference of the genius personalities is the ability to simultaneously operate with nine or eleven ideas. In any case, the ability of a person to operate with many objects is limited to a small number. And it significantly determines the possible extent of various human actions (the number of subordinates of the manager, the dimension of models easy to use, etc.).

Group peculiarities. *Clanthink*. This term refers to a situation where all members of a group believe in the same false statement. Historical examples are past periods when everyone believed that the earth was flat and that bloodletting was a universal remedy; or that the sun orbited the earth. The only way to overcome clan-think is to create conditions under which the group will be convinced of the incorrectness of the generally accepted view.

Groupthink. This is indeed a pathological condition of the group, in which its members hold a common opinion despite the obvious and well-known facts, clearly contradicting it. Group thinking manifests itself in the following forms:

- members of the group feel strong discomfort if a dissident objects to the general opinion;
- pressure of the group deprives the individual of independence of thought, encourages him to unanimity; critical thinking is suppressed;
- the norm of behavior is the demonstration of loyalty even with the obvious infidelity of the policy and the undesirability of its consequences, despite remorse;
- double standards: different attitudes toward "ours" and "aliens"; the humane and noble attitude to ours, and irrational cruelty to others; the disregard of ethics and morality for achieving objectives (e.g., the atomic bombing of Japanese cities by the Americans);
- voluntary abandonment of doubts in the name of consensus;
- illusion of own invulnerability, "super-optimism", despite the chain of failures and unlucky (e.g., Kennedy's action in the Bay of Pigs, Johnson in Vietnam, NASA in the history of the disaster "Challenger");
- the illusion of unanimity ("Silence is a sign of consent"); consensus is considered proof of correctness.

The main means of overcoming group thinking is to create conditions for the expression of different opinions, to encourage criticism, to be open to information from the lower levels of the organization and from outside, including signals from rivals and opponents, and to take all this information into account for decision-making.

Dissent (Spreadthink). It is a fact that differentiates estimates with group members of the same problematic situation. It is explicitly scrutinized during the formation of the problem mess.

Thoughtlessness. Any group that comes together to achieve a goal can do so in an effective and productive way, or go the hard, slow way to insignificant results. The latter is a pathology.

A good example of this is the work of many commissions (committees) set up to deal with a problem. A number of circumstances (intentional and unconscious, internal and external to the commission) often result in the commission's performance being void. R. Akoff jokes quite seriously about this: "the amount of useful results produced by the committee decreases with the increase size of the committee itself. Consequently, the optimal number of members of the committee is zero. <...> The function of the committees is not to make decisions, but to delay their adoption long enough until the issue under discussion is resolved by itself. <...> The committee appreciated the higher, the more time and money it requires to not come to any conclusion, and the more voluminous is report on how they have come to this". R. Akoff sees the measure of struggle with this pathology in the fact that the one who holds the meetings of the committee must pay for the time of those who are present. Although sometimes the commission is created specifically to delay the decision.

Peculiarities of organizations. Because organizations are established structures of subordination of the hierarchical type, they are additional to the personal and group specific "pathologies" that must be overcome in the systemic transformation of the organization.

Pathologies in the hierarchy occur for two reasons

The first is that higher-ranking managers feel superior to lower-ranking employees. This feeling leads to underestimation or even rejection of any improving suggestions from subordinates. Measures to overcome this pathology are seen in the introduction in upbringing, education, training of managers, thus instilling in them the ability to support any actions of subordinates that increase the efficiency of their work.

As the situation becomes more complex, it is time for change. Including changes in the activities of the smallest groups in the organization, of which the upper layers of leadership often do not think. Such a situation needs managers (own or attracted from outside) who know how to overcome the complexity with the help of effective complicit, participative techniques, countering personal and group pathologies.

In particular, managers need to develop a sense of *receptivity to the innovative proposals of their subordinates.*

The second reason for the emergence of pathologies in the hierarchy is that the lower ones themselves feel a sense of humiliation, dependence on the higher ones. For example, during a meeting, participants almost always wait for the chief executive to speak. After that, the speakers try not to contradict this. Because of this, competitive alternatives are not considered. This may be countered by giving the floor to the manager at the end of the discussion, and not for making a decision, but for making additional options for further discussion of all proposals.

Hence, the reasons for organizational pathologies in the hierarchy lie in the fact that from the top is the fear of losing control, and from the bottom, the fear of doing something wrong.

The result: inefficiency, stifling creativity.

Antidote is the movement toward participatory management: "What do we want to achieve?" "How can we help each other?".

Practice shows that both these pathologies are surmountable by a method of the systems analysis, except one — arrogance and rejection from the top management. The only way to remedy this situation is to replace managers with people who are aware of the role of pathologies, including their own.

Another type of organizational pathology is related to the way the organization treats the mistakes of the performers.

To do something right is only to confirm what is already known; you cannot learn from it. But on the mistakes you can learn if they are recognized and corrected. But in most organizations, mistakes are frowned upon and even punished. This blocks development.

There are two types of errors. The *mistakes of committing* are to do what should not have been done. The *mistakes of noncommitting* are not to do what has to be done.

Errors of noncommitting are more serious than errors of committing because (besides other reasons) they are often impossible or very difficult to correct. These are lost opportunities that will never be returned. Organizations often fail because they have not done something, rather than because of what they have done (e.g., the Communist party of the USSR).

However, the errors of omission are rarely recorded and even more rarely punished. Therefore, the perpetrators are punished for doing something that should not have been done, but they are not responsible for not doing what should have been done. Since only mistakes of committing are punished, the optimal strategy of a manager seeking his own safety is to get rid of them, up to doing nothing. The most successful leaders are those who create the appearance of activity without doing anything. This is the root of the organization's reluctance to change.

Making change outside requires change in yourself. In Chapter 3, we repeatedly stressed that any activity of the subject is carried out using suitable models that provide this activity with the necessary information.

The models needed at the moment are extracted from the set of innate and acquired by the subject models forming his "world of models" and "culture" of the subject. Thus, the success of the activity and the quality of the result significantly depend on the culture of the subject. This, of course, applies to management activities, including solving problems, that is, interfering in reality and changing the behavior of people involved in the problem situation.

Changing the behavior of the subject is difficult, even for the individual subject, even if the change is a matter of life or death. For example, studies have found that after cardiac bypass surgery, only 12% of patients change their habits to a healthier lifestyle, reducing weight and exercising. And what then to speak about difficulties of change of behavior of the subject — organization consisting of many individuals!

Progress in this matter has led to research in the field of human psychology, which has revealed the mental characteristics of the subjects, encouraging them to resist one form of managerial influence on the part of the head and obey others. The conclusions can be formulated as follows [1].

Change is painful. Why are people so resistant to change even if it is in their own interest? One reason is that changes are required not only in the environment but also in the individual's models system (i.e., culture). (In fact, the resistance to changes is a universal law of the nature: inertia in physical world, conservation laws for matter and energy, instinct of self-preservation in living beings, conservatism in human societies.)

After a long enough practice of driving a car, people begin to drive by "not thinking": new models that provide it, go down to the subconscious level of actions in an organization. Once in a country with left-hand traffic, a person suddenly experiences serious difficulties driving a car. Some even refuse to drive in a foreign country. A similar situation arises in the transition from the control of the machine with an automatic gearbox to manual control.

The same stressful situation arises in any strategic or organizational changes: there is a need to move from the usual, proven working procedures of interaction of workers to new, unusual ones requiring increased mental stress. Because of a purely animal instinct that causes anxiety when expectations and reality diverge, many people experience stress and discomfort of various forces and do everything to avoid changes. Managers making changes in the organization should take into account that the same events are perceived differently by people at different levels of the organizational hierarchy, that is, they experience different stresses on this change.

Behaviorism doesn't always work. The conditioned reflexes discovered by Pavlov on dogs were generalized by B. Skinner in the 1930s to people in the form of behaviorism, the theory of organization governance on the basis of rewards and punishments ("stick and carrot", "whip and honey-cake"). This technique is widely used both in the routine management practices and in the implementation of updating changes.

However, psychologists have found that signals designed to excite the desired response through a conditioned reflex do not always work unambiguously for people. For example, a manager may reprimand those who are always late for meetings. This may shame them and have a temporary effect, but it shifts their attention from the essence of the matter to the problems that cause delays. Another manager may prefer to reward in some form those who show up on time. But at the same time, late people experience unpleasant, disturbing emotions that do not contribute to the cause.

Humanism is overrated. In the 1950s and 1960s, humanistic tendencies focusing on personality, emotional needs, and values became a major priority in psychology. This psychology prescribed assistance in realization of personal potential through self-actualization and manifestation of hidden opportunities and aspirations. From the practice of encouragement and punishment they moved to empathy, complicity, and insight into the personal problems of subjects and the transition to the practice of holistic solutions to problems. This is done by creating in a group relation of trust and mutual understanding, awareness of the need, and value of change.

The instinctive desire of all living organisms to maintain balance and evasion of change (homeostasis) is one of the reasons for resistance to any changes. An effective way to overcome this resistance is the involvement of stakeholders in the design of changes for themselves, the organization of this process by a system analyst with the help of special techniques (facilitating), helping participants to develop their own decisions.

The focus on anticipating change and encouraging feedback is crucial. An example of how expectations create reality is the placebo effect: assure the patient that he has taken an analgesic, and he will actually feel relieved, although the pill was completely neutral. The main thing is to shift attention from the problem to its solution, to encourage people to look for this solution themselves, to focus their attention on the creative side of these searches. The method of idealized design proposed by R. Akoff [4] is the most consistent implementation of these ideas.

The role of ethics in systems analysis. The above paragraph is partly a free paraphrase of the "theory of systems practice" by R.L. Ackoff, an outstanding American systems analyst, late Emeritus Professor at the University of Pennsylvania [10–15].

It is interesting to mention another section of the theory of practice of R. Ackoff, dedicated to ethics and morality, categories that are inevitable in any activity and in the systems analysis, too.

On the one hand, systems research has much in common with "conventional" scientific research. The performer must be conscientious, honest, objective, faithful to the truth, demanding to his competence, observe the norms of communication with colleagues in the profession.

On the other hand, in the systems analysis, in addition to the actual truths ("objective", "scientific"), it is necessary to take into account many subjective factors: personal human values, psychological aspects of relations between people, individual assessments of reality, etc. These factors are poorly studied, far from strict formalization, and extremely specific for each person, which significantly increases the importance of ethical aspects in the behavior of the systems analyst.

For example, one of the dangers ("pitfalls") in systems analysis is that a systems analyst imposes his or her own opinion on stakeholders, including the decision-maker. The ethics of a systems analyst's behavior is not to be a "gray cardinal," that is:

- not to hide alternatives that for some reason he does not like himself; to bring such alternatives to the attention of the decision-maker;
- the same with regard to alternatives that may or may not even be liked by the decision-maker;
- explicitly communicate the assumptions underlying the findings;
- to draw the attention of the decision-maker to the sustainability or sensitivity of alternatives to changes in conditions.

A special issue is the inevitability of compromises and their limits. The idea of improving intervention can also be expressed in this way: the aim is not to seek the truth, not to find out who is right and who is wrong, but to reach agreement among all.

The implementation of improving intervention will inevitably require compromises. A simple example is when a customer insists on including a part that he or she considers essential in the working model, and the analyst has the opposite opinion. To create a favorable psychological atmosphere, the analyst must agree with the client, although it may be a mistake of the first kind.

However, compromises are not always so painless and so permissible. A systems analyst is faced with an ethical choice when his principles contradict the principles of the customer. Some ethical rules for a systems analyst in such a situation have been proposed by Dror:

- not to work for the client who does not give access to the necessary information;
- do not perform analysis only to justify an already made decision;
- not to work for a client whose goals and values contradict humanistic values and analyst's own beliefs.

The categorical nature of these rules in reality sometimes encounters so-called "complexities of life". Ethics is generally not a forced matter, but a voluntary one.

For example, the well-known cyberneticist S. Beer carried out systems research on the problems of managing Chilean economy by order of the Allende government, but refused to work at the invitation of Pinochet, although he then had to take personal security measures.

However, at an international conference on systems analysis, the view was expressed that such an uncompromising attitude should not be absolutized. The argument was this: the customer knows that his and your ethical attitudes are contradictory. Turning to you, he thus shows a willingness to change his attitudes. Why not use the opportunity to improve customer ethics? "Imagine that the priests would refuse to deal with sinners; then the Church would have nothing to do." The speaker gave an example of solving the problem of racketeering of small shopkeepers by hiring their guards from among the racketeers.

Of course, my sympathies are on the side of S. Beer, and not those who offered to involve the bandits themselves in solving the problem of racketeering. But what will I do if criminals come to me and offer me a choice: either I help them solve their problem for a big reward, or they will terrorize my family? I am craving for having enough moral strength to follow the example of Stafford Beer.

Questions and Tasks

1. What does it mean to "assess risks" of the project?
2. What kind of control is called "management"?
3. What are the reasons (three) for which the participation of stakeholders in the systems analysis is necessary.
4. Why should we seek voluntary participation of stakeholders in the analysis?
5. List the three conditions that ensure voluntary participation.
6. What are the measures to meet these conditions?
7. Try to name the ethical norms of scientific research, in general, and applied systems analysis, in particular.

Part II References

1. M. C. Jackson. *Systems Thinking: Creative Holism for Managers*. John Wiley & Sons, Ltd.: University of Hall, 2003.
2. R. L. Ackoff. *Re-Creating the Corporation*. New York: Oxford University Press, 1999.
3. J. N. Warfield. *Introduction to Systems Science*. Singapore: World Scientific, 2006.
4. R. L. Ackoff, J. Magidson, H. J. Addison. *Idealized Design: Creating an Organization's Future*. Wharton: University of Pennsylvania, 2006.
5. Ф. И. Перегудов, Ф. П. Тарасенко. *Основы системного анализа*. 3-е изд. Томск: Издательство НТЛ, 2001. (F.I. Peregoudov, F. P. Tarasenko. *Fundamentals of Systems Analysis*. 3rd edition, Tomsk: NTL, 2001.)
6. Д. А. Поспелов. *Большие системы. Ситуационное управление*. М.: Знание, 1975. (D. A. Pospelov. *Large Systems, Situational Governance*. Moscow: Znanie, 1975).
7. Н. Г. Загоруйко, В. Н. Ёлкина, Г. С. Лбов. *Алгоритмы обнаружения эмпи рических за кономерностей*. Новосибирск: Наука, 1985. (N. G. Zagoruyko, V. N. Elkina, G. C, Lbov. *Algorithms for Detection of Empirical Regularities*. Novosibirsk: Nauka, 1985).
8. R. L. Ackoff, H. J. Addison. *Management f – Laws. How Organizations Really Work*. UK: Triarchy Press Ltd, 2007.
9. J. S. Hammond, R. l. Keeney, H. Raiffa. *Equivalent exchange: a rational approach to compromise //* Effective decision-making. Series "Harvard Business Review", 2006.
10. R. L. Ackoff, E. Vergar. *Creativity in Problem Solving and Planning*. In Eupopean Journal of Opreational Research, 1981, № 7, pp. 1–13.
11. E. de Bono. *Simplicity*. London: Penguin, 1999.
12. R. L. Ackoff. Theory of Practice in Social Systems Sciences. In: *Art and Science of Systems Practice. Proc. of Roundtable at IIASA, Nov. 6–8,!986,* Laksenburg, Austria.
13. R. L. Ackoff. *Ackoff's Best: His Classical Writings on Management*. NY: Wiley & Sons, Inc., 1999.
14. D. Rick, J. Schwartz. *The Neuroscience of Leadership. Strategy + Business*. (www.bah.com). Reprint № 06207, 2007.
15. R. L. Ackoff. Magidson, H. J. Addison. *Idealized Design*. Wharton, 2006.

Part II References

Part III

Brief Review of Results of Systemology in the 20th Century

6 The Current Stage of Development of Systems Thinking

The Transition from the Ideology of the Machine Age to the Ideology of the Systems Age

6.1 INITIAL IDEAS ABOUT THE STRUCTURE OF THE UNIVERSE

For a successful action in the real world, it is first necessary to correctly (adequately) imagine what it is like. It turns out that the world in which we have happened to be born is arranged far from simple. From the interactions with the world around us, humanity has understood that:

a. the world is **material**, and we have the concept of ***matter***, substance, and the corresponding laws of its preservation (according to M. Lomonosov, "If somewhere something decreases, then somewhere else the same will increase as much");
b. the world is **heterogeneous**, and we have a concept of separate *objects* and their sizes and distances between them, that is, the concept of ***space***;
c. the world is **constantly changing**, and we have the concepts of ***movement***, ***energy***, and ***time***, with the corresponding conservation laws (efficiency cannot be more than 100%, and time is irreversible);
d. the world is **structured**, and we have the concepts of ***organization*** and ***information***; more generally, the notion of ***systems***, with their various degrees of development, including our own system, called the brain (in our opinion, the most developed).

Each of these four aspects of the universe discovered by humans is manifested in reality in countlessly diverse forms. Even the very concept of the materiality of the world in recent years has been replenished with the idea of theoretical physicists about the existence of "dark" matter; hence, the matter we sensed turns out to be only an insignificant part of the matter of the universe.

The variability of the world also appears in various forms. In trying to understand these forms, we introduced the concepts of periodic and nonperiodic, fast and slow, deterministic and random processes. Changes in our environment occur at an ever-increasing rate, while their unpredictability increases. The speed of processes that are very important for our existence (e.g., movement in space, performing computational operations, or establishing personal connections with another distant subject) in the time of our generation increased by orders of magnitude more than during the entire previous millennial history of mankind. In many ways, this was one of the reasons for the need to develop new, more efficient methods of dealing with the processes occurring with and around us.

Especially important for our existence in the world is the *degree* of knowledge and understanding of what is happening around us, and the awareness of the possibilities to influence it. Here comes to the fore the fourth aspect of reality — information interactions. The key point is the fact that changes in our environment can occur not only because of natural patterns but also as a result of our subjective intervention in the course of events. What we consider necessary to do, as well as do, is determined by our vision of the world (information about its present state), our goals (information about its desired state), and our ability to make the transition from the first to the second (information about available resources for action). The idea that allowed us to bring together our ideas about the infinite qualitative-spatial-temporal diversity of the universe as an infinite set of distinguishable entities was the concept of a system.

We can now think of the reality surrounding us as an unlimited set of interconnected and interactive systems, neighboring (in time and space) or containing smaller systems, or entering as parts into larger systems.

But perhaps no less important in the systemic view of reality is the realization that the properties of the system are not just a total set of the properties of the parts, but, fundamentally, qualitatively new, inherent only in the system as a whole, quality. The properties of water are not just a mixture of the properties of hydrogen and oxygen; the properties of the useful substance — table salt—are not a combination of the properties of toxic substances, sodium and chlorine; the properties of a family are not limited only to the properties of the husband and the wife; no part of the aircraft flies, but only the plane flies as a whole; etc. Any such property of the system, which is not reducible to the properties of its parts and not derived from them, is usually called *emergent* (in the static version) or *synergistic* (in the dynamic).

6.2 THE PECULIARITY OF THE HUMAN SYSTEM: THE CULTURE OF THE SUBJECT AS "SECOND NATURE"

The development of system concepts led us to an understanding of the fundamental difference between man and other objects of reality. To emphasize this difference, the object "person" received even a special name — the *subject.* The subject has the peculiarity that it interacts with the environment not only passively following the unshakable laws of nature, like all other natural objects, but also actively transforming the environment in accordance with its subjective goals and in the degree accessible to it (see Section 2.4 in Part I). The subject is both an individual and a group of people united by a common goal, which can be of any size, ranging from family, tribe, organization, ethnic group, to the state and humanity as a whole.

The purely subjective interactions of the subject with the environment are its *cognitive* and *transformative* activities. Both of these activities are carried out through *modeling* — human activity on the accumulation, transformation, and usage of information.

Cognition is modeling the existing environment; *design* is the construction of models of a nonexistent, but desirable state of the environment; *governance* is an action aimed at eliminating the discrepancies between the existing and desired states by a gradual and purposeful change of reality.

The combination of all models of the subject (and innate, through which begins the development of the newborn subject, and acquired ones as a result of its own experience of existence in a particular environment) forms *its culture*, which determines the entire character of the subject's behavior. Various actions of subjects in the same external conditions are associated with differences in their cultures, which are purely individual. The similarity of their behavior is associated with the presence of common elements in their cultures, and conflicts between subjects are caused by differences in their cultures.

Therefore, we have to talk about the individual, unique culture of each subject — person, group, organization, nation, etc. This is very important to understanding the phenomenon of man. All researchers of human nature have necessarily paid attention to the crucial role of the culture of the subject in its life. They gave it different names that emphasize certain aspects of it: a model of the world, a picture of the world, a mental model, a thing for us, an inner world, ideology, tacit knowledge, semantic space, paradigm, beliefs, a priori assumptions, pattern, stereotype, perception filter, frame, installation, anticipatory scheme, conceptual structure, context, latent knowledge, etc.

An important extension of the meaning of the term "*culture of the subject*" in relation to the above terms is that the concept of the subject's culture includes not only abstract models in the human brain, that is, mental models in the *spiritual* culture of the subject, but also real, material models that constitute the *material culture* of the subject, including his senses and all artificial systems created by him: tools, works of art, household items, appliances, equipment, buildings and structures, and in general, the infrastructure of human life. The entire culture of the subject determines both the perception of reality and the planning of actions, as well as their implementation in the interaction of the subject with the environment. Since any interaction of the *subject* with *reality* occurs through culture and by culture, reality is presented for us in our culture, we tend to believe that nature is what we "see" in our models. Therefore, culture is sometimes called "*second nature*". (This feature of our thinking was noticed by Saint Augustine[1]: "Miracles do not contradict nature: they contradict what we know about nature"!).

[1] Aurelius Augustine, or Blessed Augustine (354–430) — the most influential preacher, Christian theologian, and politician. Holy one of Catholic and Orthodox churches. One of the Fathers of the Church and the ancestor of the Christian philosophy of history.

The process of *governance* is based on the comparison of the real and desired (target) states of reality, and the success of management depends on the quality of our knowledge of reality (cognitive *models*!), as well as on the consistency of our goals (pragmatic *models*!) to the laws of nature. In the simplest case of managing artificial (in particular, *technical*) systems, the condition for meeting these requirements is Ashby's "law of necessary diversity": the system can be completely successfully managed only if the controlling system is no less complex (diverse) than the controlled system. The necessity of this requirement follows from the fact that for fully successful control the control system must "know" (model) all possible states of the controlled system.

6.3 THE DEVELOPMENT OF THE MODEL OF THE UNIVERSE: A PARADIGM SHIFT

Until recently, the desire to successfully manage not only technical but also social systems, in which people are essential elements, was to ensure and satisfy Ashby's law, that is, the controlled system was simple and acted like a machine, so that people in it behaved like parts of a machine and performed only those functions that are necessary for the operation of the entire machine. The division of labor, the mechanization of production processes, the commission of people to perform those elementary operations that (so far) are not amenable to automation, in fact, completely dehumanized the work of people on conveyor lines, continuous productions, in factories and plants, and in rigid hierarchical organizations. The entire educational system that has developed over the past centuries has been subordinated to the goal of training personnel for work in *analytically* separated areas of labor specializations. In other words, the organization of society was based on the *mechanistic paradigm*, on *analytical thinking*, on the preference of only causal relationships between any entities that interest us, and on the organization of social structures by simplifying the system's hierarchical relations of subordination between its elements.

Although historical epochs do not begin and end with specific dates, but replace each other gradually, the dominance of the mechanistic paradigm is associated with a specific period. Russell Ackoff believed that the Machine Age began during the Renaissance and ended during World War II. The peculiarity of the Machine Age is that in the management of social systems, the treatment of people was based on the fact that they, like the details in the mechanism or internal organs of a living individual, should perform only the functions assigned to them and should not manifest their own goals. Two circumstances have so far ensured the success of these models (mechanistic and organismic) of society.

First, the achievements of an analytical, cause-and-effect approach to the study and modeling of any manifestations of reality were impressive: the division of the complex into smaller and yet smaller parts made it possible to explain how this complex is arranged and how it operates (see Section 3.2 in Part II). Almost all our areas of knowledge are organized analytically: the analysis is carried out to find the elements that constitute all the objects in this area. In physics, these are elementary particles; in chemistry, atoms and molecules; in biology, cells; in music, notes; in linguistics, the analysis is brought either to individual words, or to symbols of written speech, or to phonemes in oral speaking; in management, analysis generates hierarchical executive structures; etc.

Second, during the Machine Age, people were forced to accept the fact that employers (heads of enterprises and organizations) did not take into account the personal interests and goals of employees: it was difficult to find a job, and only on conditions dictated by the employer. The qualifications for performing simple operations were of a low level, and managers gave instructions to all their subordinates regarding how to perform these operations.

Over time, the situation began to change. It became increasingly clear that nature is much more complicated than any of its models, that universal interconnectedness in nature makes abstraction an idea of the causal connection of only two entities, and, most importantly, that only an analytical approach cannot provide comprehensive explanations of reality (especially, human nature).

On the other hand, the development and sophistication of technologies required an ever-increasing skill of the workers, and as a result, workers began to understand their affairs more and better than their superiors. Consequently, the old methods of managing organizations began to lose their effectiveness. It has become increasingly necessary to take into account that the management of the social system must be based not only on the objectives of the system itself but must also take into account the interests of large (social and environmental) systems, of which our system is only a part, and in addition (and obligatorily!) take into account the personal characteristics and interests of each employee.

Under these conditions, it became necessary to increase the controllability of social systems, approaching the Ashby law of maximum controllability, not only by simplifying (reducing the diversity) the controlled system but primarily by increasing the complexity and development (increasing diversity) of the controlling system. The mechanistic paradigm began to be supplanted by the systems paradigm: The Age of Systems is coming.

A new world view is a new culture, dictating the promotion of other goals, and the creation of other ways of behavior in general and management in particular. Global changes are associated with the development of our understanding of the truly incomprehensible complexity of reality, and with the transition to taking into account the individual characteristics and interests of each subject.

If the term "*analysis*" is used in the sense of "ascertaining how the system is composed and how its parts interact", and the term "*synthesis*" in the sense of "ascertaining how the system interacts with other systems from the environment", then the new systemic paradigm differs from the old "analytical" one, complemented by its sharp increase in attention to the synthetic consideration of the system itself and all its parts (considered not as elements but also as systems).

Such a paradigm was formed due to the efforts of many system thinkers and practitioners of the 20th century, among which the most notable are R. Ackoff, L. Bertalanffy, A. Bogdanov, T. de Chardin, E. Deming, P. Drucker, J. Forrester, D. Meadows, N. Moiseev, D. Pospelov, I. Prigogine, P. Senge, V. Vernadsky, and N. Wiener.

The main feature of the new paradigm is a significant step toward improving the adequacy of our models of reality: now, to achieve our goals, it becomes fundamentally important not to be limited only to information about the internal structure of the system, which we intend to transfer into the desired state, but above all to take into account the possible consequences of our planned changes in the surrounding system environment. This does not mean abandoning the *analytical* description of the situation, but supplementing it with a *synthetic* consideration (see Section 3.2 in Part I).

There is another important point in the ongoing development of the new paradigm. Building the adequacy of our models cannot be limited only to taking into account more and more information about the world around us. The need was realized to include in our working models the reality and information about the nature of the person himself, who creates the models.

We now know that the function of modeling our brain is not limited to *analysis* and *synthesis*, which are the results of the work of our consciousness on the logical transformations of the available information. It turns out that very many and very important forms of modeling, ensuring the vital activity of a person, occur at the *subconscious* levels of the brain.

The study of subtle characteristics of human behavior (psychology, sociology) and the functioning of the human brain itself (anatomy and physiology of the brain) revealed the existence of special and diverse processes of subconscious (implicit, latent, nonverbal) information processing in the brain, generating amazing, sometimes even seeming to us supernatural, models: those that, although to varying degrees, are manifested by everyone (instincts, emotions, intuition, creativity, dreams, insight), and those that do not manifest in everyone and not always (hypnotizers, geniuses, prophets, clairvoyants, psychics, telepaths), and those that are considered unhealthful deviations from the norm (pathological mental diseases).

Some results of the study of these nonconscious information processes are used in the practice of social systems governance. For example, hypnosis is used for medical (and sometimes political) purposes; modern pedagogy sets the task of early identification of innate abilities and talents in children for their subsequent conscious maximum disclosure. In modern management theory, the problem of consciously involving the irrational intuition of managers in their rational management decisions is intensively developed (this becomes critical when there is a shortage of necessary information and time for finding a rational solution; and such situations are not uncommon in management practice).

The system picture of the world, which is the basis of the mentality of mankind, develops not only in the form of a change in the most common paradigms, as mentioned above. The development of systemology is also occurring, extracting from the real world all the new experimental information about it (*data mining*) and transforming this "raw" data into the new elements of our system models (modeling technology, *data processing*, *knowledge management*). Consequently, our understanding of the nature of systems expands and deepens, which allows us to plan more successful changes to our environment to realize our goals.

Different goals require different information about the system and its environment in various combinations of information about their different properties — static, dynamic, synthetic (see Chapter 2 in Part I). The development of the models of the corresponding groups of system properties has reached a level where it is possible to speak about the formation of specific sections of systemology — *systems statics*, *systems dynamics*, and *complexity theory*. Let us briefly describe the current state of these sections.

Questions and Tasks

1. Which features of the real world do we reflect in terms of matter, space, time, and system?
2. Try to show that modeling is not a function that a person can do or not do, and that he cannot do anything at all without modeling.
3. Formulate the main differences between the mechanistic and systemic paradigms (in the vision of effective relationships between people in the process of their life in the environment).

7 Elements of Systems Statics

When systematically considering a particular problem situation, one has to appropriately combine information regarding these properties with data on its other qualities. When using this information for practical purposes, it is important to keep in mind that *all properties of the system are not independent, and are interrelated.* Working with the system, you must constantly keep in mind that it is integral, and that all its properties are not manifested separately, but jointly and simultaneously: all the properties of the system themselves also form a system. Let us discuss the static properties of systems taking into account this aspect.

7.1 INTEGRITY (COMBINED WITH OPENNESS, FUNCTIONALITY, EXPEDIENCY, AND EMERGENCE)

The idea of the *integrity* of the system reflects the fact that the real world is heterogeneous, and that one heterogeneity can be distinguished from another as a separate *whole* — by some features, characteristics, and properties that are inherent in this heterogeneity but not in others. We perceive something as a whole, if it is the owner, the carrier of some specific *property.*

The property that distinguishes one object from others is thought of by us as an attribute of an object that belongs to it. The ball is round, the weight is heavy, the grass is green, the girl is beautiful, etc. However, on closer examination, the concept of *property* turns out to be the product of an amazing feature of language modeling — the ability to simply, effectively, and economically describe a complex situation, *replacing the multi-place relation with a single-placed one.*

Any property of an object becomes known to us only by observing its interaction with other objects; it is a description of the peculiarities of its *ties* and *relations* with them. For example, when we say that "glass is transparent", it means that "if a light source is placed on one side of the glass and a device sensitive to light on the other, the device will register the passage of light energy through the glass", and you can even measure (using a special technology!) the degree of "transparency" of the glass. Consider another example, we say that "this object is green". This phrase briefly expresses the fact that "when this object is illuminated with white light, the light analyzer, sensitive to colors, will find that this object reflects only the green component of the entire spectrum of white light frequencies".

Such an understanding of "property" refers *to all properties of all objects*: any property of an object manifests itself only in its interaction with other objects. The explanation of what a described object is, is reduced to listing its properties of interest to us, that is, some of its connections with other objects. Which of the properties you need to describe depends on the purpose of the consideration of the object. The presence of system interactions with the environment is expressed by the concept of *openness* of the system. At the same time, Kant drew attention to the fact that the external properties of an object, which become known to us by observing and understanding its interactions with its environment (cognition "things for us"), are a manifestation of some internal features of the object that are inherent in it, regardless of the state of the environment and the presence of an observer. Kant calls such features "a thing in itself".

Transparency of glass is manifested (glass becomes "a thing for us") not only when it is illuminated with light, but also in darkness, a glass (as "a thing in itself") has the property that makes it transparent when illuminated. This "something" is knowable only in some part of it, in the course of a certain specific experiment, and in the absolute sense remains unknowable. For example, we do not know what people are capable of and to what extent until we test them in action. The modern development of pedagogy sets the main goal of education (upbringing and training) of a person to detect innate (*internal*, potential, purely individual) abilities and talents of a person, as well as to create conditions for their full disclosure in the *external* environment. The revealing of a previously unknown property of an object is called *discovery*.

On the other hand, the integrity of the system can also be viewed as the *internal* appearance of emergence. Violation of integrity (e.g., the removal of any, even the smallest, part of it) changes the entire system (see Section 2.3, Part I). Ecologist A. Leopold [1] puts it figuratively: "If the soil mechanisms work well as a single whole, then every part is good, and it doesn't matter how detailed we are in understanding them. For millions of years, biota managed to build something that we really like, but in which we understand almost nothing, so getting rid of parts that we think are useless to us — is a monstrous stupidity". Although he referred to soil, this is true for any natural system (e.g., at one time there was a practice in the United States to cut an appendix from all newborns to prevent appendicitis in the future; of course, the practice had to be stopped, although the role of the appendix in the body remained unclear). With artificial systems, the situation is the same: by altering them, we assume the role that natural selection plays in nature.

7.2 OPENNESS (COMBINED WITH FEASIBILITY AND FUNCTIONALITY)

The openness of the system is displayed by its "black box" model. The main difficulty in modeling reality stems from the fact that everything in the world is interconnected and interdependent, and each time it is necessary to include in its black box model, from the entire infinite interweaving of connections of our system with the rest of the world, only a very limited *finite* number of connections, information about which is necessary and sufficient (*essential*) to solve *our* particular problem. Therefore, of the various connections, we include only "strong", "important" inputs and outputs in the model. At the same time, because of the subjectivity and the relativity of the concept of essentiality (in relation to the achievement of a goal), difficulties arise in the form of the possibility of making four types of errors (discussed in Section 2.1, Part I).

Analytical description of the system's relations with the environment begins with consideration of significant *pairwise* interactions between the system and each of the elements of the environment. At the same time, it is necessary to take into account that relationships that differ among themselves *qualitatively* in nature and *quantitatively* in the strength of one entity's influence on another, and, naturally, we strive to take into account first of all the most "essential" (for our purposes) in our understanding.

The strongest of them, *cause-and-effect*, are called the laws of nature. The causal connection between events is so strong that it is ***necessary and sufficient***: if the first has happened, the second will necessarily occur, and vice versa, that is, if the second has happened, then the first has necessarily preceded it.

The presence of third parties is not required in the interaction for a causal relationship (Figure 7.1). However, this implies that such a "law of nature" is a certain abstraction from reality with its *universal* interconnectedness. This approximation manifests itself both in practice and in theory. For example, independence of the acceleration of free fall from the mass of the body follows from the law of gravity; but stone and fluff fall differently because of their interactions with another entity, *air*. In theory, it manifests itself in the form of paradoxes. For example, according to Coulomb's law, the force of interaction of two charges tends to infinity as they approach each other; according to Newton's law, the strength of mutual gravitation of the two masses increases indefinitely with decreasing distance between them (in both laws, the force of interaction is inversely proportional to the square of the distance).

Thus, the causal relationship between the two entities is an *approximate* description of reality when other interactions are neglected. For example, when designing a crane, only the weights of its parts and cargo (i.e., the forces of their interactions with the earth) are taken into account, neglecting the (very small, but existing) forces of their mutual attraction to each other.

However, in many situations, the difference in significance of different influences on the system is not so great that it is permissible to neglect some of them; when describing the interaction between only two entities, the interference of the others in this interaction should be taken into account. This intervention may have a different character, which leads to three more types of presentation of the final interaction as a *pair*, weaker than cause-and-effect.

1. The connection of two entities may be ***necessary, but not sufficient***. An example is the connection between an acorn and an oak: for an oak to appear, an acorn is necessary, though not enough — soil, moisture, heat, light, air, and much more are needed, without which the oak will not grow. A term for this type of connection between the two entities has not yet

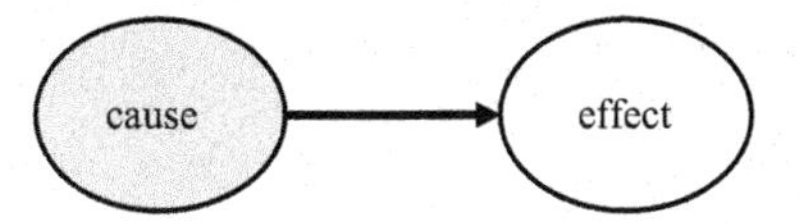

FIGURE 7.1 Causal relationship abstracts from universal interconnectedness.

been established. In the group of causal conditions, the principal cause (the essence of the acorn) is called the "*main cause*" or "*system-forming factor*", the remaining (also necessary) conditions are called "*accompanying causes*" or "*limiting factors*". R. Ackoff proposed to call the central pair of entities so connected (the main cause-and-effect) as "*producer–product*" (see Figure 7.2).

Producer is a "system-forming" factor, and without it, the remaining ("limiting") factors are useless; however, without the limiting factors, the producer cannot affect the product. Consideration of system dynamics, that is, its history after the origin, led to the expansion of the concept of this type of communication. In addition to the root cause, without which the consequence cannot occur in principle, many other factors are necessary for the functioning of the system. Importantly, their presence is required in strictly defined proportions between themselves and in strictly defined limits for each one. For the growth of oak, soil is necessary, but not of any composition; it needs moisture, but in a certain proportion; it needs heat, but in a limited range of temperatures; etc. If any factor goes beyond the permissible limits, it will cause the failure of the entire system (like a drought or a flood ruining the harvest). Hence, it becomes a factor *limiting* the effect. Often, the actual management of the system is reduced to returning to the normal limits of limiting factors that succeed each other (due to changes in time and the environment, and the system itself) throughout the life of the system.

Thus, this type of relationship is characterized as *the relationship of the effect to the causal combination of system-forming (fundamental) and limiting factors.*

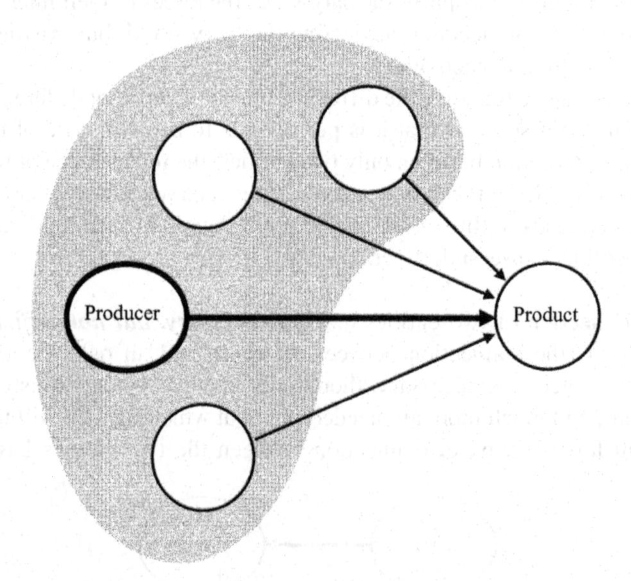

FIGURE 7.2 Among the factors defining the result, there are the main one and the accompanying others.

2. The connection of two entities may be ***sufficient, but not necessary***. This is the case when the same effect can arise from any of the various causes (like being wounded in a battle from different attacking objects; or intoxication from different drinks; or getting the required amount from a number of investors). Such a connection is sometimes called directed or potentially possible (see Figure 7.3). Each of the several factors leads to the same result.
3. The connection of two entities can be ***significant, but not necessary and not sufficient***. Unlike the first three, *direct* interactions between two entities, they can be linked *indirectly* through chains of other interconnected entities. There may be several such chains, and the effects on one chain may be positive, while on others it may be negative (see Figure 7.4). The final result for different objects in the same circumstances may be different, and when it is considered as a whole, the set of objects is interpreted as *probable, possible, random.*

An example of the indirect link between two entities is the relationship between smoking and a smoker's heart condition. Smoking affects the heart through effects on several various parts of the body: it tightens the arteries (which is bad for the heart), stabilizes body weight (which is good), reduces anxiety (which is good), damages the lungs (which is bad for the heart), etc.

Such an indirect connection may have different strengths, and in some cases becomes significant. At the same time, by virtue of final ambiguity and randomness, indirect connections *allow one to predict events, but they do not have explanatory power.* According to Ackoff and Emery [2], "Even a very strong link between alcoholism and socio-economic conditions does not explain the causes of the disease and does not help prevent or cure it. Statistics of road accidents and the state of intoxication of drivers does not help to eliminate accidents on the roads". Relations of this type are called *indirect*: "*stochastic*", "*probabilistic*", "*statistical*", or "*associative*".

Distinguishing between different types of relationships is important primarily because different relationships require a different approach to modeling and/or

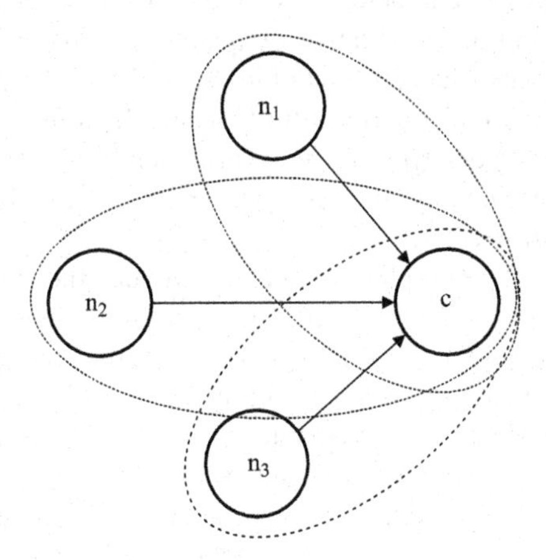

FIGURE 7.3 Case when any of the factors may produce an equal effect.

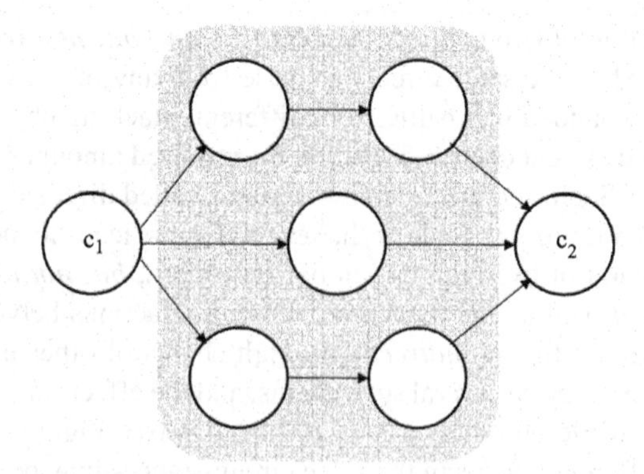

FIGURE 7.4 Indirect connections between two entities.

controlling the object. This becomes especially clear when one connection is mistaken for another, for example, "after — means due to", or an error of the fourth kind when building the black box model, or when a strong stochastic connection is taken as cause-and-effect ("in a healthy body — a healthy spirit", "smoking causes lung cancer", etc.).

7.3 DISTINCTIVENESS OF PARTS (IN COMBINATION WITH FUNCTIONALITY AND PURPOSEFULNESS)

This property is associated with the external boundary separating the system from the environment, or the boundaries between the heterogeneous sections within the system. To the discussion (in Section 2.1, Part I) about the difficulties of constructing a composition model, it is worth adding a more detailed discussion of the very concept of the boundary of a certain entity.

Whichever entity we consider, we describe it by listing a set of its properties, observed or presupposed characteristics, and attributes (measurable parameters recorded in the appropriate qualitative or quantitative *measurement scales*). The state of a particular entity (the entire set of its elements) is characterized by a certain set of values of these parameters. If we think of each feature as a coordinate of some space, we get the *state space* (*phase space*) of the entity, in which its static (instantaneous) state is represented by the corresponding point and the dynamic one (time process) by a certain trajectory.

A boundary is a surface in phase space that divides the space into two regions. Although such a partition can be carried out as you like, when it comes to determining the *boundary* of a system, the partition must meet the appropriate conditions: the limited subspace of the system states contains many different (differing in coordinates) elements but have at least one feature common to all elements of the system, which is not inherent to elements not included in it. This specific indication is called a classifier. For example, the condition of each patient is characterized by their own set of values of signs (gender, age, weight, height, temperature, blood pressure, blood composition, etc.), but there is a classification feature that allows you to divide the space of states into two areas: one includes patients with a disease, the other includes healthy (or sick with other diseases) people.

Finding boundaries between areas corresponding to different diseases is the goal of every doctor for diagnostic purposes. Often the main distinguishing feature of the system is its possession of some specific *function*. When the output process of the system is considered as its objective function, the definition of the boundaries of the system is facilitated: it includes everything involved in the implementation of this objective goal. This makes it possible to clearly distinguish between the nervous, circulatory, and digestive systems in the body; or the heating, electricity, and sewage systems in the building; or the healthcare, education, and public security systems in the state; etc.

Another reason for separating the system from its environment can be the fact that often the connections between parts of the system are much stronger than their connections with the elements of the environment, or they sharply differ qualitatively. For such systems, the boundary appears to be a natural partition between them and the environment. A striking example is the external framing of physical bodies — the surface of various objects, technical structures, organs in the body of animals, planetary systems, settlements in the territory, islands in the ocean, etc. In such cases, consider the *natural boundaries* of the system. Apparently, this is what the Father of Cybernetics N. Wiener had in mind when he spoke about information links in social systems: "The system extends to the limits to which information reaches".

However, (by virtue of the law of universal relationship in nature!) even natural boundaries sometimes have to be described roughly. For example, the atmosphere of a planet with a height is thinning very gradually, and its outer "boundary" can be carried out only conditionally. Moreover, the presence of *a subjective factor* in the relationship between reality and the subject makes the distinction relative, conditional, and subjective. Different subjective goals of interaction with this system force us to "cut out" the contour from the objective goals of the intertwining of all and with all.

For example, the boundaries of a university look different when viewed for different purposes. In the land cadaster of the city for the University is assigned a certain territory. In labor law, the belonging of the employee to the university is defined situationally (remember the example with the payment of sick-list in Section 2.1, Part I). With distant learning, students of the university may be located in remote regions.

Another example is the river. Geographical maps of the world only display the riverbed. A geophysical approach to the definition of the river system includes all tributaries and catchment areas, extending the concept to the river basin. Environmentalists consider the boundary of the river system as its watershed with the basin of another river. From the positions of the river fleet, river boundaries narrow from the banks to the limits of the fairway, designated by special signs (buoys).

Another example is man. In society, external border of humans is "judged by their clothes". The visible natural outer border is the skin. Doctors consider the outer boundary of the body as the shell of the gastrointestinal tract, relating its contents to the external environment. Some researchers of the phenomenon of man have established the presence of a special field surrounding the human body (the Kirlianu effect, biofield, aura), which is its integral component.

Hence, the main factor in carrying out the boundaries of the system and its parts is the *subjective goal* of building a model of the system's composition. This led to one of the definitions of the system as a set of interacting parts: *the system is all that we want to consider as a system*. The objective basis for such a statement is the universal relationship and interdependence in nature, that is, *the integrity of nature*: nature itself is a single whole system of systems.

7.4 STRUCTUREDNESS (IN COMBINATION WITH THE INTERNAL HETEROGENEITY OF THE SYSTEM, ITS OPENNESS, FUNCTIONALITY, EMERGENCE, AND PURPOSEFULNESS)

The model of the system structure is defined as *a list of significant links between parts of the system*. The keyword here is the evaluative term "*essential*". This concept includes the following:

1. information about this relationship is *necessary* for successfully achieving the goal with the help of the created model;
2. the *information content* of this relationship far exceeds the importance of other relationships recognized as "insignificant" and therefore not included in the model;
3. the aggregate information about all significant relationships is *sufficient* for the successful achievement of the goal.

In fact, the model of the structure arises as a result of purposeful refinement of the composition model: for each selected part of the system, a black box model is built, the inputs and outputs of which are significant connections of this part with others.

It is appropriate to start by considering the simplest (at first glance) element of the structure model — a separate *tie* between two fixed elements of the composition model. An important feature of the relationship is often its *direction*, reflecting the fact that the interaction of two entities includes their effects on each other. There are situations when the action of one on the other is equal to the action of the second on the first (as in mechanics — "action is equal to counteraction"); however, often one of the effects is superior in strength to the other, and then in the working model of the situation remains only the most powerful (as in the model of relations between the boss and the subordinate, or in the model of relations between cause and effect). Consequently, the system model often looks like a graph with unidirectional (*causal*) connections between pairs of vertices.

In many cases, the nature of the dynamic processes in the system depends on the kind of resource, with what intensity comes through this communication channel, and what are the possibilities of regulating this intensity. Then, there is a need to provide information about the *composition* and *structure* of the *connection* itself, and to make appropriate changes and additions to the structural scheme of the entire system. Within this approach, the functioning of the system is represented as the flow of resources (substances, energy, information) through channels connecting parts of the system, each of which carries out certain (quantitative and/or qualitative) transformations of the resources received as inputs, and sends the results to other parts through output channels. At the same time, there are three typical components of the model of the composition of the connection (the first is inherent in any connection, the other two are present only in some connections):

- the communication channel ("flow", "pipe", "pathway", etc.), the maximum possible flow rate of the resource through the channel is called its "capacity";
- flow regulator ("valve", "tap"), which allows you to change the flow rate in the channel between zero and the capacity;
- a reservoir ("storage pool" of a resource), in which the volume of the resource is characterized by a "stock" (or "level") and depends on the ratio between its capacity and the speed of its input and output flows.

As a result, the system composition model includes not only its own functional parts but also the specified parts of each link (see Figure 7.5). Taking them into account is very important while considering the dynamics of processes in the system. (This will be discussed in the next paragraph.)

Further, the system can be considered a sequential chain of connected parts. The action of one entity on another can be carried out not only directly but also through the impact on third entities that are associated (directly or indirectly) with both. The chains of such intermediate entities can be arbitrarily long, and the result of the impact through a particular chain can be both positive and negative. If there are several such chains between the two entities (remember the example of the effect of smoking on the heart), the final effect is determined by the predominance of one over the other.

Another, and very important, feature of the sequences of directional ties is that the chain of pairwise influences, which began from this object, can ultimately be closed to itself. This phenomenon is called *feedback*. In this case, the return effect on the original source can contribute to the initial process and strengthen it (in this case, the feedback is called *positive* or *stimulating*), but can also weaken or oppress it (and then the feedback is called *negative* or *stabilizing*). The presence of feedbacks leads to a wide variety of system behaviors, which is the subject of the theory of system dynamics. (We will discuss this issue later when considering the dynamic properties of systems.)

By considering individual links and their linear sequences, we proceed to the description of the features of the whole set of links — the model of the *structure* of the system. Together, the composition and structure models form a *structural scheme of the system*, usually represented by a *graph* consisting of "nodes" or "vertices"

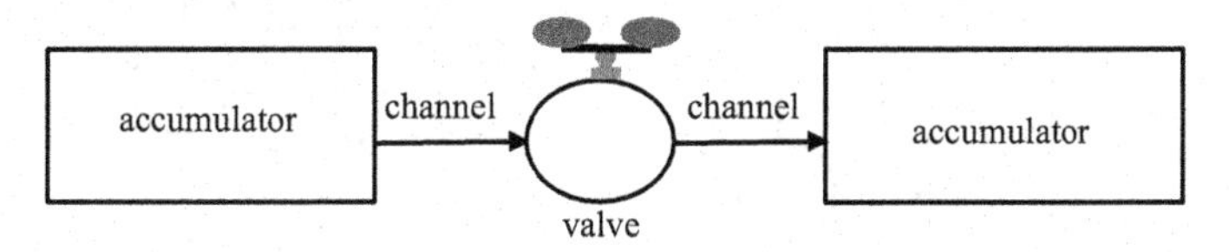

FIGURE 7.5 Elements of the resource flow.

(image of parts) and "edges" or "arcs" (image of links between parts). The direction of relationship is indicated by arrows on the edges (*oriented* graphs); the difference in the quality of the relationship is sometimes displayed by multicolored arcs (*painted*, or *colored* graphs) or other distinguishing features; for example, J. Forrester [3] displayed the flows of different types of resources in economic systems by arrows of different configurations (see Figure 7.6).

Duality of graphs, in which either parts or their functions can be considered as vertices, allows describing both the static and dynamic structural scheme of the system. This model enables the representation of a wide variety of systems. Their block diagrams form a wide range of different graphs, from a linear chain to a *complete* graph, in which each vertex is connected to all the others (Figure 7.7).

On graph models, it is convenient to consider the different features of systems related to their structure. Let's consider some variants of system structure models.

One of the aspects that provide specifics to the structural scheme of the system is the *reliability* of the system, that is, its proper functioning, despite the failure

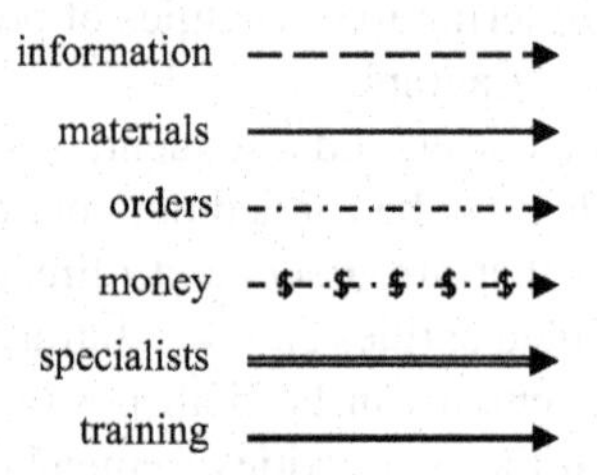

FIGURE 7.6 Flows of various resources presented by different arrows.

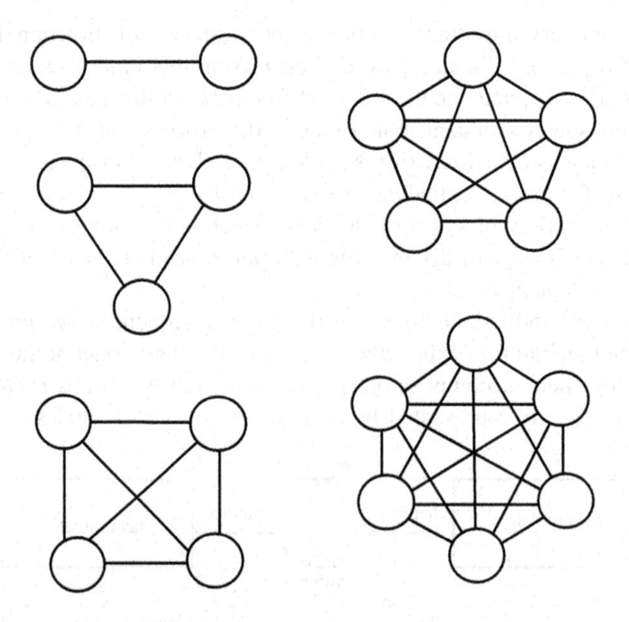

FIGURE 7.7 Examples of full graphs.

of some element(s). Increasing the reliability of the system is possible by introducing *redundancy* in the system structure. In technical systems, this is often done by duplication or parallel execution of several elements of the same function; in information transmission systems in the presence of noise, redundancy is introduced into the transmitted signal when it is encoded (the simplest example of error-correcting coding — multiple repetition of the same message); in organizational systems, the necessary redundancy is usually introduced in the form of reserve (e.g., the posts of the president and vice-president, chief executive officer and his deputy).

Hierarchy has a special place among structural schemes. This term refers to a multilevel tree-like structure diagram in which each element is associated with only one top-level element and with several lower-level elements (see Figure 5.26, Part II). This structure gives the system the properties with the results that the vast majority of natural and artificial (man-made) systems are organized hierarchically. The specific features of ideal hierarchical structure are as follows:

- the system consists of a certain set of parts, the internal connections of the elements in each of them are stronger than the external ones, that is, the boundaries between the parts are *natural*, and each part performs its *function*, that is, has its own, imputed to its *purpose*;
- the goals of the parts are agreed among themselves in such a way that their joint implementation ensures the fulfillment of the system's goal, of which they are parts. The goals of the parts in the hierarchical structure form a *tree of goals*, with each element playing a dual role: on the one hand, it is the goal in relation to the subgoals of the lower level, on the other, it is *a means* (*subgoal*) of achieving the goal of the upper-level element.

Many natural and artificial systems have a hierarchical structure because this structure allows simply building up hierarchy levels to create much more complex system of relatively simple elements, while maintaining harmonization of the objectives of the parts for the purpose of the entire system. Hierarchy is a very effective means of overcoming the complexity of large dimensions.

Perhaps the main advantage of hierarchy is that it allows the storage and constant processing of huge amounts of information necessary for the existence, growth, and development of the entire system in a changing environment, with a significant limitation that each element in the structure can only work with limited amounts of information.

For example, the number of subordinates to a manager is limited by the purely psycho-physiologic limits of a person's ability to keep in mind the activities of several subordinates at once simultaneously (empirically established in psychology rule "seven plus or minus two"). If you need to coordinate the actions of more subordinate workers, the way out is to create a hierarchy. In natural physical systems that are expanding in size, hierarchical organization is often realized in the form of *spiral* structures (shells of some mollusks; vortices, hurricanes, and tornadoes; in mollusks, whirlwinds, hurricanes, and tornadoes; the shape of the Milky Way and other galaxies in the universe).

A common variant of system hierarchy in nature is the *fractal* structure, with the same rules for combining elements at all levels of the organization (in branching plants, geological structures, in the organization of matter from elementary particles to macrocosm).

In dealing with any part of the problem, consideration of the situation may focus on the different neighborhoods of that part in the overall structure of the entire system, taking into account the different degrees of interaction of that part with other related parts. For example, if you have a disease of any organ, the doctor may prescribe treatment aimed at changing the state of only the organ, or its interaction with other subsystems of *the same level of hierarchy* (e.g., with the cardiovascular system); however, the doctor can also pay attention to the type of your psyche and lifestyle (related to *higher levels of hierarchy*), or the molecular structure of cells of the diseased organ (which is at the *lower level of the structural hierarchy*), and even gene alterations in the molecular structure of your DNA (in hereditary diseases).

Although hierarchical systems evolve from the lowest level, and the original purpose of the upper levels of the hierarchy is to help the lower levels achieve their goals, in real life, for various reasons, there are deviations from the rules of the ideal hierarchy, changing the characteristics of the system as a whole. *Social* systems are characterized by deviations because people (as *goal-setting* subjects) have some goals at odds with those dictated by the rules of the hierarchy. Among such situations, the most common are the following:

- people belonging to different branches and levels of hierarchy often establish and use for their own benefit connections that are not provided by the ideal hierarchy (e.g., nepotism and collective guarantee in the authorities);
- subjects pursue not only the goals assigned to them by the hierarchy but also other goals often more diligently (e.g., if a corporation bribes power structures to lobby its interests, society suffers from the violation of the mechanism of market competition; if students believe that their goal is to get good grades, not knowledge, then general cheating begins, preventing the realization of the goal of education);
- in many social hierarchies, the upper levels tend to maximize centralization of governance. The excessive control of the center over all elements of the hierarchy leads to the fact that each of them is deprived of the opportunity to perform the functions of their own maintenance, as a result of which the entire system may perish (history gives many examples of this).

However, hierarchy is not the only possible structure of complex systems. When a system becomes more complex, the distribution of functions across its components can be less asymmetric and centralized, and the importance of parts for the system as a whole becomes less diverse. In such cases, *multidimensional* and *network* structures are formed.

This is especially true of *social* systems in which the essential elements are people with their own subjective goals, which is contradictory to the requirements of the ideal hierarchy. The network is qualitatively different from a hierarchical organization: there is no part of it whose removal would disrupt the functioning of the entire system. For example, the destruction of a direct telephone line between two cities does not prevent them from communicating through a third city to which they are both connected.

Network structures in recent years have attracted increasing attention due to their increasing importance in public practice. Typical examples are marketing and logistics networks in the economy, collective information structures (social networks) on the Internet, distributed computing structures in computer science, and some mass phenomena in sociology and psychology. R. Ackoff [4] drew attention to the fact that the network nature of international terrorism makes dealing with them ineffective, based on the assumption that terrorists are members of an organization and not a network. Networks require a different approach than organizations. The study of network structures has become an actual problem in the theory and practice of management.

Questions and Tasks

1. Show using an example that the concept of property as an attribute of an object is a simplified model of its multilateral relations with other objects.
2. List four types of relationships between two events.
3. Where are the boundaries of the system?
4. Discuss the differences between the structure and the structure model of the system.

8 Elements of Systems Dynamics

The set of knowledge about the possible time processes in the world of systems and the factors that affect what kind of process will occur is the content of a special section of systemology — *system dynamics.*

Let us try to give an overview of the main results of studying system dynamics obtained to date, discussing each of the dynamic properties in conjunction with other system properties.

8.1 FUNCTIONALITY (IN CONJUNCTION WITH STRUCTURING, PURPOSEFULNESS, AND STIMULATING)

A *function* of the system is called a dynamic process that occurs at a certain *output* of the system. For example, the function of a vehicle is to move some cargo in space; the function of the lamp is to light a certain space around itself; the function of the enterprise is to produce certain products (goods or services) that satisfy a particular need of society; etc. As there are a lot of outputs (connections of this system with the environment), we're talking about the objective *multifunctionality* of any system.

The connection between functions and structure is that each *function* of the system is the cumulative result of the actions of all parts of the system, and this result depends on the *structure* of the links between the parts: it can be either the result of a nonlinear, emergent interaction of parts of the system (synergetic effect), or the manifestation of a linear total set of qualities of individual parts of the system (nonemergent properties, such as the total weight or the total volume of the structure consisting of parts of different weights and volumes).

The functionality property is directly related to the *purposeful* property. First of all, for any artificial system, some of its function is the target process for the realization of which the system is created and used. (Recall that the *subjective goal* is defined as the *entire trajectory* of the system transition from the "problem situation" to the "ultimate purpose"; see Section 2.3 in Part I.) This process is needed by the subject; but to be realized in reality, it must be an *objective* function of the system. This allows us to talk about *any* real process at the output of the system as its *objective* goal.

There are different ways to realize a subjective goal. Sometimes it is possible to find a natural system, the natural functions of which perfectly corresponds to our goal, and it remains only to organize its use (e.g., cow milk production is its natural function, but it coincides with one of our many subjective subgoals, and we include dairy cattle as an element in our artificial dairy production system; the same applies to natural sources of the natural resources we need, food and industrial, material and energy). However, in many cases, there is no system in nature that has the function we need in the form necessary to achieve our goal. For example, birds fly, but they cannot be harnessed, like horses, for our flying.

Sometimes there are no natural systems with the functions we need. And then we try to create an artificial system that carries out the process we need. The main difficulty lies in the fact that you cannot implement any process that you just imagine to be wanted. Only subjective goals that do not contradict the laws of nature, that is, can become the objective goal of the system created by you, are achievable.

Another striking manifestation of the relationship of functionality, purposefulness, and stimulation is the management process. Even after creating the necessary artificial system and launching it into action, we are faced with the fact that the system works in an ever-changing environment. All the factors involved in the operation of the system change, and some of them can go out of the range of values allowed for the normal functioning of the system. Such factors become limiting and take away the real trajectory of the system from the target trajectory. This requires intervention, measures to overcome the difficulty, and returning the system to achievement of the goal; such actions are called system governance. Different limiting factors require different actions to overcome them, which results in the presence of different types of control with specific algorithms (see Section 4.3 in Part I).

8.2 STIMULATION (IN COMBINATION WITH INTERNAL HETEROGENEITY, STRUCTURING, FUNCTIONALITY, AND PURPOSEFULNESS)

In the initial sense, *stimulation* refers to the fact that any system, among its connections with the environment, has such connections through which *the environment affects the system*, causing some changes in its state. But for the subject, this opens up a wonderful opportunity to manage events for their subjective purposes, causing the outputs of the system necessary (target) processes by specifying certain (targeted) effects on its inputs.

All human practice is based on the use of this opportunity. Any desired change in reality is carried out by us precisely by applying control actions on the inputs of the controlled system. However, the same practice shows that it is not always possible to achieve exactly the process that is needed at the output of the managed system. When it is possible to determine the cause of the difficulties, it is possible to determine the way to overcome them (such cases are described in Section 4.3, Part I, as seven types of control).

Naturally, the more complex the managed system is, the more difficult it is to manage control problems. It is not surprising that many surprises are presented to managers of social systems. Even the notion of counterintuitiveness of the behavior of such systems has taken root among theorists of complex systems governance: they often surprise us and behave quite differently from what is expected of them.

As shown in Section 4.3, Part I, trial and error is a reasonable way to manage complex systems. For the management of the social system, it is often first reduced to trial attempts to find a "key", "leverage" point in the system, the impact on which can lead to the elimination of the problem. Such points are among the controlled inputs of the system, and the problem is to determine which of these inputs is the best leverage point for solving the problem. This is where the counterintuitiveness of complex systems is manifested (especially, in the management of social systems): the impact on the "obviously key" regulated parameters of the system often leads to unexpected results. Examples of this are failures in management, failures of a number of reforms of different (including state) scale, obvious inefficiency of many laws, etc.

Of course, the main reason for this is the complexity of the controlled system (in the next chapter we will discuss the different types of complexities). However, an important role is also played by the fact that different points of influence on the system have different potential to influence the final result at the output of the system. Donella Meadows [5] considered 12 types of interventions in the system that can be made in social systems, and ordered them by increasing efficiency, the strength of changes made in the system as a result of intervention. Let us reproduce the main conclusions of this remarkable work.

By importance for management the key points, which make it possible to change the state of the system, are ordered (by the increase of the "lever" effect) as follows.

1. Quantitative, numerical indicators: variables and constants.
2. Stocks of material resources.
3. The structure of resource flows.
4. Delays in the control loop.
5. Negative feedback.
6. Positive feedback.
7. Information flow.
8. Rules of motivating people (encouragement, punishment, constraints).
9. Self-organization.
10. Objectives of the system.
11. System of views and concepts (paradigm).
12. Expansion of worldview (going beyond the paradigm).

Let us briefly describe each item from this list.

1. Quantitative parameters of resource flows. Experience with simple (technical) systems suggests that the behavior of the system can be changed in the right direction by simply turning the tap that regulates the flow of resources between some parts of the system. In the management of social systems, the vast majority of disputes about the choice of a specific impact on the system are around what values to give to the numerical parameters of resource flows. However, in complex systems (due to the multiplicity of different factors together generating a specific behavior of the system), the change in the flow of a resource in one of the channels often has little effect (and sometimes does not affect it at all) on the course of events. The increase or decrease in the number of officials does not lead to an improvement in the work of the authorities. The increase in the cost of maintaining the police or the increase in the punishment for crimes does not lead to the disappearance of crime.

It is not that the numerical parameters are not important at all: in the short term, their changes can be significant, especially for those directly affected by the regulated flow. However, the behavior of the system as a whole is almost never affected. Adjusting the numerical parameters, Donella considers the weakest way to control a complex system, and even compares it with the bustle of rearranging deck chairs on the sinking "Titanic" in the hope to stop the sinking of the ship. The only condition under which the numerical parameter of the resource flow can become a key point of impact on a complex system is that a change in this parameter should trigger some of the more significant factors in the following list.

2. The capacity of resource storage and the amount of accumulated reserves. Sometimes the problem is to maintain a stable system operation with sharp fluctuations in input and output resource flows. Then the key point is the availability of a reservoir for the accumulation of resources and the amount of reserves in it. The accumulation of resources allows not to reduce production during the decline in the consumption of its product, and vice versa, the presence of reserves allows not to reduce the consumption of the product while reducing its income.

The resource drive plays a stabilizing role and softens the impact of fluctuations in resource flows on the system; therefore, the drive is sometimes called a "buffer". Therefore, warehouses and tanks are created in industrial, commercial, transport, energy, hydraulic, and financial systems. To abandon the cost of construction and maintenance of reservoirs can only be possible when ensuring the stability of the flow of resources, as is the case with the implementation of the "*just in time inventory*" system in construction, when the necessary parts of the erected structure arrive at the construction site just at the moment of their mounting. Storage capacity is a key parameter of the system in a changing environment, but this parameter is difficult to change: usually, the reservoir is a bulky physical structure (river dam, warehouse building). Therefore, this parameter is set only at the time of design and creation of the entire system, and if such a need arises later in the operation process, it is necessary to rebuild the system physically (as, e.g., in the construction of car parks in large cities). Therefore, buffers in the list of keypoints are important for strategic management but unpromising for operative management.

3. The structure of resource flows. The operation of the system is ensured by transporting the necessary resources from one part of the system to another. The nature of the system's functioning depends heavily on how the process of transporting resources between all parts of the system is organized. Of great importance is the proper implementation of each individual flow between a particular pair of parts (determining the required capacity of each channel, the ability to control the flow rate in it, etc.). However, no less important is *the structure of the entire network* of channels.

In Hungary, the road network is created so that when you travel from one part of the country to another, you have to pass through the capital city of Budapest; this has created many problems that are not solved only by the regulation of traffic. It is no accident that in recent years, long-distance routes in Russia are built with a detour of settlements. The Rocky Mountain Institute (United States) has achieved a surprisingly strong reduction in energy consumption in the buildings of the Institute only by changing the structure of the heating system (they straightened extra bends in pipelines and changed the cross sections of some pipes). The efficiency of many management institutions is significantly increased by optimizing the structure of the document flow in them.

The structure of flows in a system can strongly affect the behavior of the system, and therefore is related to leverage points. However, because the change in the physical structure of the network of flows of material and energy resources is almost always associated with capital construction, it is preferable to take this factor into account at the design stage of the entire system.

4. Lag: delayed response of the controlled and controlling systems. All real systems are inertial: upon receipt of some influence on the system input, the formation of the corresponding new output state occurs only after a certain time, the duration of which is determined by the inertia of the system ("*time constant*", "*coming into being period*", "*lag time*", "*delay*"). This fact is of particular importance in the management processes, the whole point of which is to develop by a controlling system impacts on the managed system in response to changes in the managed system and/or in the environment. It is obvious that the result of the control will depend on the ratio of the speed of the changes and the reaction rates of both systems.

The variety of final processes in the control system is very large: the slightest changes in the ratio of reaction rates can lead to a change in the nature of the entire dynamics of the process. The solution of dynamic equations of the system in the theory of automatic control demonstrates this variety, covering options from smooth processes, various damped and increasing oscillations, to changes catastrophically destructive.

Sometimes even without mathematical analysis it is clear what the consequences of delays in the control loop are. It is clear that a controlled system cannot respond to changes that are not commensurate with its time constant.

For example, in the 70s of the last century, the academician V.M. Glushkov fought for the computerization of the management of the economy of the USSR, considering the main reason for the difficulties of centralized planning of national economy, the delay in the feedback circuit: The State Planning Commission managed to form a cross-sectoral balance of production and consumption in the country of production only with a delay of three to four years. Another example is the electric power system of any country. The power plant is under construction for several years, and it gives power to the design capacity for several decades. With such a delay, it is impossible to track fluctuations in the current electricity demand. Therefore, the energy sector of any country in the world experiences long-term fluctuations between its overloading and underloading.

Delays at all stages of the management cycle are powerful key points. Here the main problem is the correct direction and magnitude of the change in delays. However, there are systems in which it is very difficult or even impossible to change the value of the delay. For example, you cannot accelerate a child's development, or the growth of the forest in the logging of the timber industry, either the recovery of populations of commercial fish, etc. It is difficult to accelerate the construction of large facilities, the development of new drugs, the training of highly qualified specialists, bringing innovative ideas to commercialization, etc.

5. Negative feedbacks. We have already established that the processes at the outputs of the system are determined by the effects on its inputs and its structure. To get rid of the undesirable behavior of the system and get the desired process at the output, we are looking for a leverage (key) point in the structure of the system, the impact on which will provide the target result. Such elements of the structure include feedback loops. In many situations, it is the feedbacks that prove to be powerful keypoints of impact.

Their mechanism of action is that the signal sent by some element on its output channel to another element is forwarded further to other elements via the network of channels, and in the network structure, there is such a chain of links, which, starting from the first element, eventually forms a cycle closed on it (hence, the name *feedback*).

In this case, two options are possible: (1) the returned signal came in the same polarity ("in phase") as the sent one, and when they are added, the initial change will be strengthened (this option is called *positive* (*reinforcing feedback*); (2) the returned signal arrived in the opposite polarity ("in the opposite phase"), and when it is added to the original change will be weakened (the name of this is *negative (balancing) feedback*). Both options can be used for management — of course, for different purposes.

Negative feedback is used when you want to keep the system in a given target state or on a given target trajectory despite external factors that diverge the system from a given target. Negative feedback suppresses these deviations, which is why it is sometimes called "stabilizing", "balancing", or "dampening". The algorithm for using negative feedback for control is described in Section 4.3, Part I; the device implementing this algorithm is called a *regulator.* Leverage force can be both the creation of a missing feedback loop, and the modification of some operations in the existing regulator.

Negative feedback is a powerful keypoint and is present in many artificial and natural systems. Examples include various automatic control systems (thermostat, refrigerator, current or voltage stabilizers, autopilot, Watt's regulator, etc.); sometimes the role of the regulator is played by a person (driver, pilot, machine operator), or an entire organization (legislative body, executive authorities), or a specially created system of relations in society (market economy, election of authorities); feedbacks play a huge role in the existence of living organisms, ensuring their viability in changing external conditions (adaptation systems, immune system, conditioned reflexes).

When choosing negative feedback as a keypoint for desired changes in a complex system, it is necessary to keep in mind a number of related features and difficulties in realizing this possibility.

First, regulation can compensate deviations from the target trajectory only within certain limits. Deviations can be so strong and fast that regulation simply cannot cope with them. Open the windows in the room and the air conditioner will not be able to maintain the preset room temperature. Hit the car on the ice and nothing will keep it on the road. Revolt the crowd and the police will not be able to restore order. The rate of logging, which exceeds the rate of forest recovery, leads to the loss of forests.

Second, we should not forget about the counterintuitiveness of complex systems. Often, even correctly found leverage point they begin to stimulate in the wrong direction. As an example, Donella Meadows [5] often cites erroneous government measures to regulate finances (prices, taxes, subsidies) in a market economy.

Third, there can be many different feedback loops in the system, and by stimulating one of them, you should consider how this will affect others. For example, some cycles most of the time are dormant, like emergency systems in nuclear power plants, fire extinguishing facilities in buildings, or our ability to sweat or tremble for regulating our body temperature. A serious mistake is to get rid of such emergency mechanisms because of their high cost and rare need. In a short period of time, the harm from this is not visible, but in the long run, we dramatically reduce the viability of the system.

As successful examples of strengthening negative feedbacks in social systems, D. Meadows cites preventive medicine, physical culture, high-quality nutrition, monitoring of controlled processes, taxes on environmental pollution, adequate penalties for offenses, etc.

6. Positive feedback. The result of the positive feedback is a sharp increase of the emerging trend in the behavior of the system. The more people become ill with an infectious disease, the more people become infected with it, this is how epidemics develop. The more money in your account, the more will run up the amount of interest. The more fast neutrons are in the radioactive mass of matter, the more nuclei they break into new neutrons, so the chain reaction develops, until the explosion of an atomic reactor or a nuclear bomb.

Positive feedback is consciously used as a keypoint in cases where it is necessary to bring the system into a nonequilibrium, unstable state, or even destroy it altogether. If you do not limit the positive feedback, the system will "go haywire" and destroy themselves (e.g., explosive weapons; the collapse of social systems, from families to empires; extinction of entire species of animals and plants).

You can limit its action in different ways. One way is to connect negative feedback to the process. The result depends on what you want to achieve and how you connect the negative feedback. Sometimes it is necessary to block the action of positive communication at all by applying a stronger negative one (such as vaccinations and blocking the channels of infection can stop the epidemic; or anticrisis measures in management; introduction of a progressive tax; or pressing the brakes in a dangerously accelerating car).

In some cases, it is desirable not to completely eliminate the effect of positive feedback and maintain some nonequilibrium state of the system. Then the negative feedback is introduced in such a proportion to the positive that the system is constantly kept in the desired nonequilibrium state (e.g., self-oscillating systems, oscillators; turbines of power plants; motors, engines; cardiovascular system of animals). Another way is to weaken the positive link itself (e.g., not to press the brakes, but to reset the gas; to reduce the growth rate of the population and economy; to limit the rate of consumption of renewable resources).

Of particular interest is the case when positive feedback leads the system into a chaotic state — violent, unpredictable, unique, but still flowing according its internal laws. This issue will be discussed later in the section on types of temporary changes in the system.

7. Information flows. The keypoints are those elements of the system, the impact on which leads to the desired changes in the behavior of the system. It is clear that different keypoints may correspond to different problems. Above, we considered the possibility of using as keypoints of different elements of the *physical* structure of the system: the numerical parameters, the volume of storage of material resources, channels of transportation of resources, various elements of the configuration of links in the network of these channels, etc.

However, in addition to the flow of physical resources, *information* flows play an important role in the systems, and the more complex the system, the greater is the role of information, becoming dominant in *social* systems. Therefore, the elements of the *information infrastructure* of the system are often powerful leverage points of influence on the system.

In full analogy with the physical points, in the information network, leverage point can be any of its elements, and the problem is to determine which of them is key to our specific situation, the supply of information in memory, or the capacity of a communication channel, or the structure of the network communication channels. Often the most powerful points of influence are *information feedbacks*.

In Holland, there is a block of identical houses, the only difference is that the electric meter in some houses is installed in the basement, while in others in the hallway, where residents could constantly see when they spend more energy, and when they spend less. In homes with a meter in sight, electricity consumption was 30% less.

This is a vivid example of a very sensitive leverage point in the information structure of the system. Here the parameter is not corrected, and the existing connection is not amplified or weakened, but a new loop of *information* negative feedback is created.

Another illustrative example is the requirement of the U.S. government (1986) that every factory polluting the air with harmful smoke must publish annual reports about it. From that moment on, everyone could know exactly what was erupting from the factory pipes in the city. There was no ban on emissions, no fines, no "acceptable" emission standards; it was only informing the public. But by 1990, emissions had fallen by 40%. One chemical company included in the list of "top ten pollutants" reduced its emissions by 90%, just to get out of this list.

The lack of information feedback is a very common cause of many troubles in the system. Creating an appropriate information flow is a very effective intervention, and, more simple and cheap than the alteration of the physical structure of the system. However, it is important to introduce the missing feedback in the right place and in the right form.

As an illustration can serve examples of possibilities to overcome the "Tragedy of the Commons" in fisheries and water supply. "The tragedy of communities" is the phenomenon of the depletion of renewable resources, where the total resource is consumed by all members of the community, each individually and uncontrollably. The tragedy that befell the world's fishing industry has led to the disappearance of many species of commercial fish. As a result of improvement of means of location of fish shoals in the ocean and creation of highly effective fishing nets, ships began to catch jambs almost completely. The reason for the tragedy was the lack of negative feedback from the state of the fish population to investors for increasing the capacity of the fishing fleet. Contrary to popular belief, the *price* of fish does not provide this link: the smaller the catch, the higher the price, and the more profitable it is to catch. This is positive feedback leading to disaster.

Another example is the tragedy of using a common well. It is not enough to inform all users of the well that the groundwater level is falling. This can provoke a race, the desire to pump water more than the neighbors until the water runs out. It would be more correct to set the price of water, growing the more the rate of pumping water exceeds the rate of its natural inflow.

Many people have a tendency to avoid responsibility for their decisions. Therefore, social systems often lack feedback channels; that's why these keypoints are so popular with ordinary people and so disliked by the authorities. When it is possible to persuade the authorities to open a feedback channel or to do it bypassing the authorities, there are big changes in the system. Recall Khrushchev's report on the cult of personality, Gorbachev's glasnost, Nixon's Watergate, or the publication of secret government documents by Assange, Navalny, and Snowden.

8. Norms and rules of functioning of subjects in the social system. All processes in the systems follow certain natural laws. Relations between parts of the system are realized by flows of material, energy and information resources. In natural systems, they are established by objective laws of nature. In technical systems, they are designed by the designer with an orientation to the implementation of the subjective goal, but taking into account the laws of nature. In social systems, information relations between subjects are established by the subjects themselves in the form of accepted norms and rules in society — ethical and moral norms; legal laws, rules, legal restrictions; power relations with the corresponding forms of motivation of subordinates in the management — one-time factors such as reward (perceived as pleasure or even enjoyment) or punishment (causing suffering), or factors acting constantly in the form of restrictions (compulsion to certain standards of behavior). Since the structure of relations in the system determines its emergent properties, the intervention in information relations in the social system, and the change of norms and rules of human behavior are extremely strong leverage points.

When Gorbachev declared perestroika and changed the socioeconomic rules, the country changed literally beyond recognition. However, one should keep in mind the counterintuitiveness of complex systems: it is unlikely that Gorbachev himself expected what happened. Even now, many reforms are carried out without the necessary caution.

Discussing this topic, D. Meadows, in particular, writes: "When I was explained what is the new system of world trade, the WTO, my systemic intuition sounded the alarm. In this system, rules are created by corporations and applied by corporations for the benefit of corporations. In these rules, there is virtually no connection from any other part of society. In fact, all the activities of the leadership of this organization is closed even to the press (no feedback). The WTO involves states in positive feedback loops, forcing them to compete in weakening social and environmental safeguards in order to attract investments in corporations. All this leads to the unwinding of 'success to successful' cycles until they concentrate a huge power and create a global system of centralized planning, which in the end will destroy everything".

9. Self-organization. A keypoint can be any feature of the system that can be affected, which leads to a change in the entire system. So far, we have considered the possibility of using as leverage points individual elements of the composition and structure of a particular system. But we can also consider the possibility of influencing the course of processes that occur not with a separate system only but with a whole set of systems.

A remarkable phenomenon in the world of *living* and *social* systems is the ability of a population (macrosystem, the totality of the same-type systems) to indefinitely prolong their existence, despite the limited time of the existence of each individual member of the population and changes in the conditions of existence in its environment. This phenomenon is based on two peculiarities of the population:

1. its ability to reproduce and breed its components, providing the population with the opportunity not to disappear because of the insurmountable mortality of each of its members;
2. its ability to survive in conditions unsuitable for the life of some of its individuals. This is achieved through a *diversity of qualities* of different components of the population, due to which individuals who have survived in unbearable conditions for others, and then pass on to their next generation their qualities that contribute to survival, can be reinforced in the population.

This macrosystem property in biology is called *evolution*, in agriculture *selection*, in economy *technical progress*, in politics and sociology *social development*, and in systemology *self-organization*.

Due to the stretching of generations in time, self-organization can serve as a keypoint for long-term, strategic management of the *macrosystem*. However, in the process of self-development of the macrosystem, an important role is also played by rapid, abrupt changes in its microsystems: mutations in individual chromosomes, discoveries in science, innovations in the economy, revolutions in states, inventions in technology, and transformations in organizations. Under certain conditions, these processes can also serve as keypoints for the management of the macrosystem (e.g., in the breeding work on the development of new plant varieties, or in carrying out a coup d'état).

The key points in the management of the macrosystem can be both components of self-organization: self-reproduction and natural selection. For example, in ruling over the state, it is strategically important to ensure "preservation of the people" and numerous and diverse measures for *quantitative growth* of the population. However, this is not enough for sustainable development: it is also necessary to ensure the *diversity* of microsystems in the state macrosystem.

Unlike biologists and ecologists, who understand the key importance of biological diversity at all levels of the hierarchy of living systems, and who care about the preservation of this diversity, in the field of policy, the attitude to the diversity of cultures of different subjects is blurred on the scale of tolerance, and often shifted toward intolerance. While promoting diversity, variability, and experimentation is a strategic means of achieving sustainable development, it is generally perceived in politics as an indulgence for disorder and loss of control. And the lever is being pushed in the wrong direction: cultural, social, market, and biological diversity is being diligently destroyed. Perhaps the reason for this lies in the fact that every culture has a built-in belief in the superiority of their culture over others. But with the dominance of a single culture, the stability of the system inevitably decreases. And we find ourselves witnessing and participating in the historical process of the gradual extinction of monocultural social systems, and the simultaneous growth and strengthening of systems that proclaim the equal value and equality of everyone — communities tolerant to the difference of cultures of all subjects.

10. Goals and goal-setting. The keypoints are the various features of the system, affecting which, you can change the output state of the entire system in the desired direction. In other words, it is *controllable inputs* to the system. Potential and force of influence of each point on the made changes is varied, and in our description of leverage points, they are listed in the order of increase in their "power". The potential of impacts on physical "points" is weaker than that of information "points". But all of them are used for one thing: to ensure that the desired process is realized at the output of the system, the sequence of desired states of the system, culminating in the achievement of the final desired state. The description (model) of this process is the *subjective purpose of the system* (see Section 2.3 of Part I, the purposefulness property).

Quite often it turns out that the dissatisfaction with the current course of events is caused not by deviations from the target trajectory but by the fact that the goal itself has become unsatisfactory. In this case, the most powerful leverage point becomes the *goal* of the system, and the control action is the process of changing the goal, that is, goal-setting. The remaining points — physical reserves and flows, feedback loops, information flows, and even self-organizing behavior — will be used to implement the new goal of the system.

As in the management through other points, in the process of goal-setting (development of a new goal), wrong decisions can be made resulting in the wrong goal. There are several reasons for this.

First, it may be that the proposed subjective goal is in principle unattainable because it contradicts the laws of nature. A technological example is the goal of creating a perpetual motion machine; and in the history of social systems, because of their enormous complexity, many ideal purposes have been utopian.

Second, since the world is an infinite hierarchical set of interacting systems, subsystems, and supersystems, each of which has its own goals, the whole set of their goals forms a hierarchical tree of goals, in which each fragment is both the goal of the system and the subgoal of the supersystem, of which the system is a part. As a result, the target-setting subject may make a mistake in the floor of the tree, and determine instead of a new target, the *means* of achieving the real goal, its subgoal.

The damage from this error is aggravated by the fact that it ignores other subgoals, without which the true goal cannot be realized. The failure of many innovations associated with this situation (forcing scientists in Soviet times to engage themselves in implementation of their research results into industry; the recent proclamation of the purpose of development of the Russian economy — modernization of five high-tech industries, without reference to specific sectors of the economy and without attention to other necessary subgoals; the renaming of militia in politia; introduction of fashionable clothes in the army; etc.).

Third, the stable functioning of the system is provided by the necessary coordination of the actions of the subsystems: each subsystem plays a certain role and realizes its intended purpose, one of the many subgoals of the global objective of the system. In addition, the intensity of the performed subfunction should be within certain limits that correspond to the successful functioning of the entire system. However, in living, and especially in social systems, subsystems have their own goals that go beyond what is necessary for the system, and, of course, seek to implement them in the first place. This introduces instability into the system, which should be taken into account when developing a new goal of the system (and new subgoals for subsystems). For example, the primary, initial goal of living and social populations is the goal of survival carried out by the continuous population growth. The goal of corporations is to increase their share of the world market without limitations (with the help of the WTO); the goal of cancer is to grow, pushing aside and replacing healthy cells.

Managing the achievement of the system's goal should include measures to maintain a balance between all the goals of subsystems (with rigid hierarchical management in social systems, the "vertical of power", there is a departure from the system of accounting for the own goals of parts of the system).

11. Paradigm (General picture of the world). Any human activity is based on the use of models containing the information necessary for that activity (see Section 3.1 in Part I).

To manage technical systems, there is enough information only from those particular areas of knowledge that were used in the creation of these devices, and often even a very small part of this information is enough. For example, to control a car, a driver does not need to know everything that its designers needed to know: only the information necessary for successful driving is needed. Complex household appliances (TV, PC, gadgetry, etc.) are controlled even by children.

However, as the managed system and management objectives become more complex, the amount of knowledge required increases, and when it comes to the governance of social and environmental systems, it requires information not only about the system but also about its natural and artificial environment, as well as (most importantly!) about the attitude of the managing subject himself both to system and to the environment, including *to other subjects.* In other words, we not only need models of how reality works, but we also need models of problem formulation and models of ways to solve them. This part of the subject's culture, the basic (for management) picture of the world, is called a *paradigm.*

The paradigm is primarily a set of convictions of the subject about how the world works. Most of these convictions are based on beliefs and subjective assumptions (Egyptian pyramids, religious tombs, and sacrificial structures of ancient peoples were built because people believed in the afterlife; skyscrapers are built because we believe that the land in the city center is extremely expensive). Some beliefs are refuted by science, and then a paradigm shift occurs (Copernicus and Kepler destroyed the belief that the earth was the center of the Universe, Galileo that earth is stationary, Darwin the faith in the divine origin and final perfection of man, Einstein destroyed the belief about the different nature of matter and energy, and Marx the faith in the immateriality of the nature of power).

The paradigm also includes generally accepted, historically established beliefs in the community of subjects, beliefs about the norms of the relationships between them: the concepts of good and evil, beautiful and ugly, civil rights and morals. Some of them are the result of a purely conditional social agreement (e.g., the main difference between different ideologies lies in the belief which one of possible attitudes to other, different from you subjects, is "correct"). Some of these beliefs are based on experience (moral and ethical standards are based on the belief, supported by practice, that their observance is good for society and violation is evil).

The paradigm determines everything in the created system and dictates what should be the goals of the system, its composition, and structure (storage and flow of material resources, feedbacks, information channels). The potential of the paradigm as a keypoint is huge: when the paradigm changes, the system is completely transformed. What is particularly important is that problems that cannot be solved within one paradigm can only be solved by moving to another.

But, although the paradigm shift does not require physical, financial, and even time resources, the subjects — individuals and societies — resist the challenges of their paradigm more than any other changes: it is a radical change of the subject's culture. Society often gets rid of those who encroach on its traditional paradigm (Jesus Christ, Giordano Bruno, Martin Luther King, etc.). Section 2.1 describes the ongoing process of paradigm shift in management. The transition from one paradigm to another takes place in other spheres of human culture. In the field of education, the technocratic approach, which focuses on mass training of professional personnel for society, is replaced by a humanistic approach, focusing on the disclosure of the individual potential of innate abilities and talents of each person. The evolution of the political paradigm is the drift of socioeconomic structures of societies from authoritarian to liberal democratic systems. The process of changing the paradigms of society is slow and difficult as society becomes aware of the practical evidence of the ineffectiveness of the old paradigm.

12. Expanding world view: Going beyond any paradigm. An even stronger lever of influence on the system than changing attitudes and beliefs is the realization that paradigms themselves are merely *models of the world order. No theory* can claim to be absolutely correct: everything we know about the world is in fact only a small, extremely limited part of information about an infinitely diverse and amazing universe. Any paradigm is effective (adequate, allows you to successfully achieve the goal) only with a certain combination of the goal of action and objective conditions for its implementation. Therefore, of all the paradigms, one should choose the one that is most suitable for achieving the goal *under the given conditions.*

For example, the principle of unity of command is the most effective way to collectively solve problems in war and emergency situations; the principle of priority of group interests is the most effective way to achieve superiority of one group over others; the principle of priority of everyone demonstrates its advantages in the growth and development of satisfaction of all five needs of human communities — in the production of goods (economy), knowledge (education and science), governance (power, politics), pleasures (aesthetics in human physical and emotional actions), values in interpersonal relations (morals and ethics). However, questions remain as to whether the pursued *subjective* objective itself is in conformity with the *objective* laws of nature and society (see Section 1.4, Part I).

8.3 SYSTEM VARIABILITY OVER TIME (IN COMBINATION WITH INTERNAL HETEROGENEITY, STRUCTURING, FUNCTIONING, AND DEVELOPMENT OF THE SYSTEM)

A discussion of the dynamics of processes at the inputs and outputs of the system is a consideration of changes occurring *outside* the system, although directly on its borders with the environment. However, in many situations of interactions with the system, the subject needs to know also what is happening *inside* the system, "how the system works". This is especially characteristic of those situations of control when the design of the control action relies on how exactly the interaction between parts of the system will occur as a result of impact. This third section of system dynamics is devoted to this topic.

The development of a management decision is aimed at finding such an impact on the system that will lead to the desired response, that is, to the implementation of the desired process at the output of the system. In this formulation of the question, it comes down to finding a "leverage" point of influence on the system and trying to predict what kind of impact on this keypoint will lead to the desired behavior of the system. This usually requires information about what processes take place in the system, that is, information contained in static and dynamic models of the composition and structure of the controlled system.

For artificial, "hard" (and, in particular, technical) systems, such information is presented in the form of a complete structural scheme (combining models of composition and structure), the technical documentation that is created during the design and construction of the system. Whatever the problem may arise in such a system, in its structural scheme, we can find the part of the information that is needed to solve this problem.

With "soft" (especially biological, social, and environmental) systems, the situation is quite different: all information about the system is embodied only in the system itself, and often the part of the information that will be needed in this situation will have to be extracted from the system itself. If it is necessary to solve the problem, it is necessary to build the necessary models based on the results of direct study of this problem situation. Technical details of the entire process of practical problem-solving are described in Part II of the book; here we will focus only on the construction of *models of the functioning* of a complex system. These models allow us to move forward in identifying the causes of the problem, which is a very important step toward its solution.

Modeling begins with building a model of the *composition* of the problem situation, making a list of all significant participants (*stakeholders*, see Section 5.3 in Part II) of the situation in question.

For illustration, we will provide such lists for imaginary situations (in real situations, much more details should be taken into account). The doctor, diagnosing the patient, considers not only the painful part of the body but also the condition of the patient's other organs, and the factors of his habitat (lifestyle, living conditions). Businessmen in solving problems take into account not only the work of their enterprise but also the factors of the external environment (suppliers and consumers, competitors in the product market, the current legislation) and employee interests.

Subsequently, follow the efforts to build a model of the structure of the situation, the definition of essential relationships between the identified factors. The construction of the graph begins, whose vertices are the factors and the arcs are the relationships between them. Further (if necessary), this static model is "revived": the direction of the influence of one factor on the other is determined (on the arcs are marked arrows, forming a directed graph); the nature of the influence is sometimes indicated only by "+" or "–" near the arrow, depending on whether the first factor contributes to or hinders the second, or the nature of the influence is determined more specifically by the words ("determines", "increases the probability", "good/bad for", etc.), such graphs are called *labeled* or *signed*; if you want to take into account the differences between the resources flowing in different arcs, they are displayed in different colors (*colored* graphs appear).

The difficulty of managing complex systems stems from the fact that the impact of some elements on other elements located in the remote part of the structural network is the result of changes in the flows flowing through several branches of the network, where each branch connecting them consists of a number of other intermediate elements. The constructed graph helps to understand the mechanism of such influence. For example, the effects of smoking on heart health can be shown (approximately, of course) as in Figure 8.1. It explains why the harm of smoking affects to varying degrees on the heart problems of different smokers: the balance of intermediate factors for each organism is individual.

However, the main difficulty in predicting the consequences of intervention in a complex system does not arise even because of the multiplicity and branching effects of parts on each other. Of great importance for how the processes in the system will evolve, is how the impact will be made on the most variable element itself. In most networks, there are not only "direct" chains of interactions but also "reverse" chains, according to which the impact returns to the initial point in a few steps and affects it, which triggers the next iteration of the impact on the behavior of the system.

The recurrent influence may be amplifying the original, and then such feedback is called *positive*, or *reinforcing*, but it can also counteract, weaken, and compensate for the initial change, and then it is called *negative*, or *balancing*, feedback. Positive feedback rapidly increases the deviation of the system from a stable state, bringing it up to disaster, while negative feedback tends to return the system to equilibrium. When there are several loops of negative and positive feedbacks in the structure of the system, the behavior of the system for us, committed to linear cause-and-effect

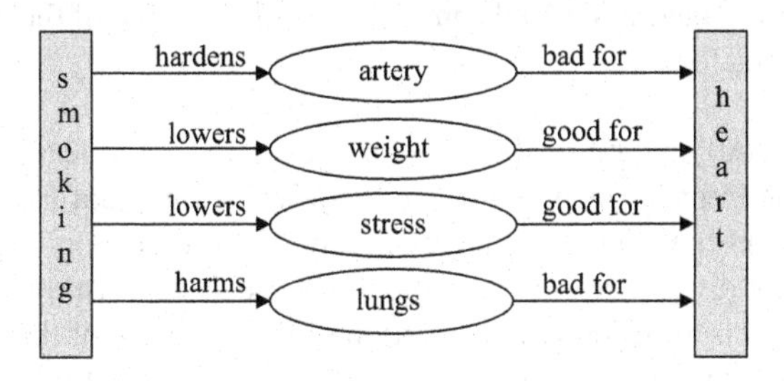

FIGURE 8.1 Effects of smoking on the heart.

relationships, becomes unexpectedly complex and incomprehensible. In addition, the complexity of the situation, the difference between the response of the system to the control action from the "intuitively expected", is greatly exacerbated by another factor in the dynamics of the system — the presence of *delays* and *lags* between incentives and responses in different parts of the system.

Due to the difference in the time intervals between exposure and reaction to it in different components of the system, the consequences in the behavior of the system from an immediate control action are stretched over time, so we have to talk about the short-term, medium-term, and long-term results of a single act of management. As a result, the system dynamics is such a situation that the forecasts of all the consequences of interference in the real system become unreliable and managers seriously talk about the surprising *counterintuitiveness* of the behavior of complex systems.

8.4 FACTORS DETERMINING THE BEHAVIOR OF SYSTEMS

Consideration of the features of the dynamics of systems has led to the understanding that the type of behavior of the system (i.e., the specific nature of its response to external influences) is determined by a combination of many factors. The main ones are:

1. the relationship between the *subjective purpose* of external influence, *the objective purpose* of the whole system, and the *own objectives* of the parts of the system;
2. the topology of the *structure* of the system: that is, the configuration of the network of links between all parts of the system; the presence in the network of direct ties and feedback (positive and negative) and their specific combinations;
3. the correlation between the speeds of resource flows through the channels between parts of the system, the channels *capacities*, and the *amount of reserves* of resources in the storage;
4. the inertia of each part of the system (the value of the *delay* of its response to the input impact); the ratio between the *times of the delay* response to the stimulus in different parts, especially, in the controlled and controlling parts;
5. the choice of "leverage point", that is, a specific (local or global) aspect of the system, which it is decided to have an impact;
6. the choice of a specific *type of impact* on the keypoint; for example, the direction and magnitude (and sometimes configuration) of the change in the controlled factor.

Each particular combination of these factors generates a characteristic response of the system to the impact. Each factor can be described by qualitative and quantitative indicators, and the slightest changes in any of the indicators of any factor may lead to significant differences in the behavior of the system. In this regard, modern systems dynamics is aimed at the elucidation of the structure of the set of behaviors of the system (i.e., on construction of various *classifications* of the processes occurring in systems).

Interesting (and useful for at least a partial understanding of what is happening in complex systems) results are obtained when considering the effect on the behavior of the system of combinations of not all factors at the same time, but only one or two of them. In some cases, processes can be described quantitatively; in more complex situations, it is possible to give only a qualitative, fuzzy description of the general trend of the process (nevertheless, such information is useful in working with the system).

We will briefly discuss the identified types of systems functioning.

8.4.1 Flows and Stocks of Resources

The processes of transportation of resources according to the structure of the system are easier to consider than others. The main parameters of the dynamics here are the speed of movement of resources in each channel (edge of the graph), linking a certain pair of system elements (nodes of the graph), and the volume (levels) of the resource reserve accumulated in each store. The limiting parameters are the capacity of the channel (the maximum speed of the resource flow in this channel) and the maximum volume of each store.

The processes are described by linear additive equations, familiar from school times on the problem of the pool with pipes, through which water flows in and out. Difficulties arise when it is necessary to comply with certain ratios between the levels of stocks in all the drives in the system, which poses the task of coordinating the speeds of all flows (e.g., the tasks of the economy associated with the ratio of supply and demand in the market, production and distribution of goods, the inflow and outflow of labor resources of the enterprise, etc.). However, this task is not among the complex ones *if there are no feedbacks in the structure of the channel network.*

8.4.2 Qualitative Models of Complex Systems Dynamics: Archetypes of Behavior

Any purposeful interaction of the subject with the system (whether it is the study of the system, the design of a new system or the management of an existing one) is carried out with the help of models of the system behavior, that is, the description of the processes occurring in the system in the process of interaction with it.

When controlling, it is important for us to know how the specific impact of $x_i(t)$ on the i-th controlled input of the system will effect on the process $y_j(t)$ on its **j**-th output. In other cases, it is important to understand why the system behaved this way and not otherwise. What quality of information we need, about which processes, and in which elements of the system, depends on the goals of the subject. The set of information about the behavior of systems, modeling (description) of various such processes and is the content of *the system dynamics.*

When working with complex (especially social) systems, it is not always possible to bring the modeling to the construction of an accurate, rigorous, mathematical model because of the practical impossibility of taking into account all the links between all elements of the system and the environment. However, often in practice we do not need all the information about the object; we need only the part that will allow us to achieve the realization of our specific goal. A clear analogy is the modeling of the same terrain with maps for different purposes. For some purposes, enough information provided on the globe; other purposes require a more detailed display of the various features of the area: there are maps of different scales and purposes — geographical, administrative, environmental, political, geological, demographic, topographic, military, secret, etc.

This situation is inherent in any kind of human activity (both practical and cognitive), including systemic dynamics. To date, efforts to develop theoretical models of system dynamics are intensively conducted in two directions: the development of quantitative and qualitative models of system behavior. The first direction, called *synergy*, focuses on the mathematical description of regularities (*attractors*) arising in the chaotic behavior of complex nonlinear systems. The second direction focused on overcoming the "counterintuitiveness" of complex systems by finding out what features of the system *structure* cause this or that character of the *functional* behavior of the system.

Very few systems (e.g., cells multiplying by division) appear immediately in a perfect state for themselves. For most systems, the lifecycle is a period of gradual changes in their composition and structure, and the corresponding changes in their functions (properties). Changes can be very diverse: quantitative and qualitative. They can occur at different rates and have monotonous or variable trends. Often there are processes that are nonlinear (nonadditive) result of the interaction of two or more characteristics; and almost always such processes are generated as a result of the feedback action, and the variety of forms of nonlinear curves is associated with a variety of combinations of a certain number of feedback loops and (or) delays in the branches of the network of links. Different combinations of feedback loops in the system structure give rise to different types of system behavior. Each configuration of loops in the structure of the system generates a special, inherent only to it, the type of behavior of the system as a whole.

The fundamental result of the rapid development of system dynamics in the 80s of the last century was the identification of characteristic (different quantitatively, but qualitatively the same occurring in different systems) types of behavior of systems. In the synergy, these are attractors; in the qualitative theory, these are archetypes. In the latter case, it was found that some certain types of behavior correspond to the presence in the structure of the system of certain configurations of feedback loops and certain relations between the delays in the reaction of parts to the incoming effects. Such qualitatively different classes of systems behavior are called *archetypes*; to date (2014), more than a dozen archetypes have been identified and explained (this work was started by J. Forrester, and continued by J. Kennedy, M. Goodman, P. Senge, D. Medows, etc.).

We note that, in practice, pure archetypes occur only when the structure of the system is indeed dominated by the configuration of relations that generate this type of behavior. The degree of influence of this configuration on the behavior of the system as a whole can be weakened and distorted as the influence of other components of the system structure increases and the corresponding degree of dominance of this configuration decreases. The basic components of any network structure are linear chains of connections between elements and feedback loops. A single feedback loop circuit may involve a different number of elements of the system, and the relationship between each pair of elements in this circuit may have a different character (reinforcing — positive, or weakening — negative). What kind of feedback (negative or positive) forms in the end the entire circuit, consisting of several pairs of elements, depends on whether even or odd number of weakening links is in the whole chain: an odd number of them (1, 3, 5, …) formed a negative feedback, and even (0, 2, 4, …) positive feedback (Figure 8.2); in this form, the law of dialectics "Negation of negation" is manifested in this case.

In turn, the structure of the system may contain a different number of feedback loops in different combinations: the loops may be separate, may have common elements with other loops, may themselves be included as components in other loops (Figures 8.3 and 8.4), etc.

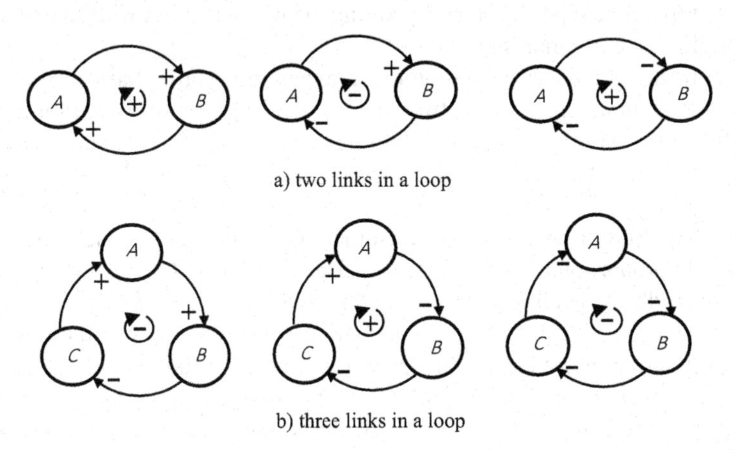

a) two links in a loop
b) three links in a loop

FIGURE 8.2 Combinations of different links in feedback.

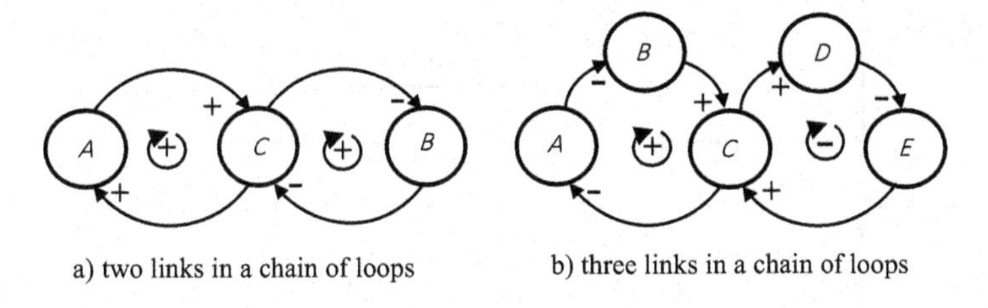

a) two links in a chain of loops
b) three links in a chain of loops

FIGURE 8.3 Combination of one feedback loop with one common element.

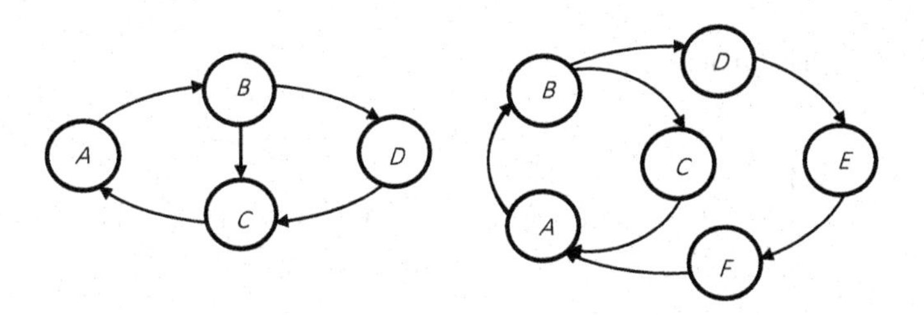

FIGURE 8.4 Combinations of two feedback loops with two common elements.

Different combinations of feedback loops in the system structure give rise to different types of system behavior. Each configuration of loops in the structure of the system generates a special, inherent only to it, the type of behavior of the system as a whole. This is how the "archetypes" of system dynamics manifest themselves.

Let us begin our consideration of archetypes with the simplest case — the predominance of **one** feedback loop in the structure of the system.

The archetype of growth of the system. Any (one-dimensional) process can be represented as a graph of the time dependence of the parameter we are interested in over time. In the most intricate curve one can distinguish "elementary", "simple" intervals of the same type and start explaining the whole process with an explanation of the origin of a particular interval.

Let us turn to the case of monotonic changes in certain characteristic of the system. For example, *linear* growth or decline of the graph is typical for a change in the stock of resource $X(t)$ in the accumulator under a constant speed difference, $v_{in} - v_{out}$ between the inflow and outflow of resource: $X(t) = (v_{in} - v_{out}) \bullet t$ (see Figure 8.5).

In the dynamics of growth of systems, there are often cases of faster than linear (i.e., directly proportional to the time) increase. The reason for this is the influence of positive feedback, leading to exponential growth (see Figure 8.6).

This (under certain conditions) occurs with the size of living population (including the population of the earth), the level of pollution of the human environment, the degree of drug dependence, epidemics, explosive chemical and nuclear reactions, etc. The account in the bank with the annual accrual of simple interest

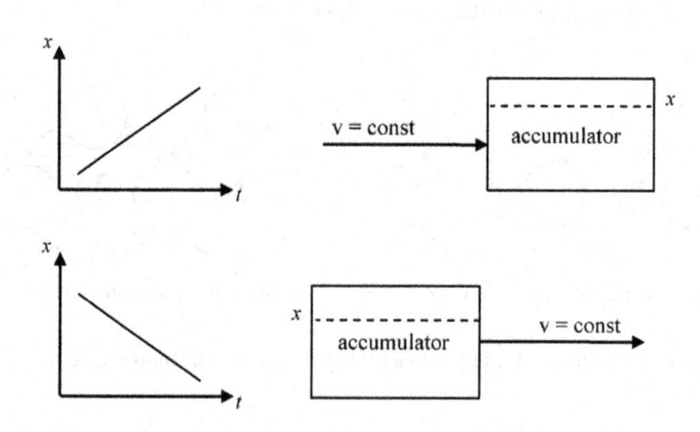

FIGURE 8.5 Linear trends.

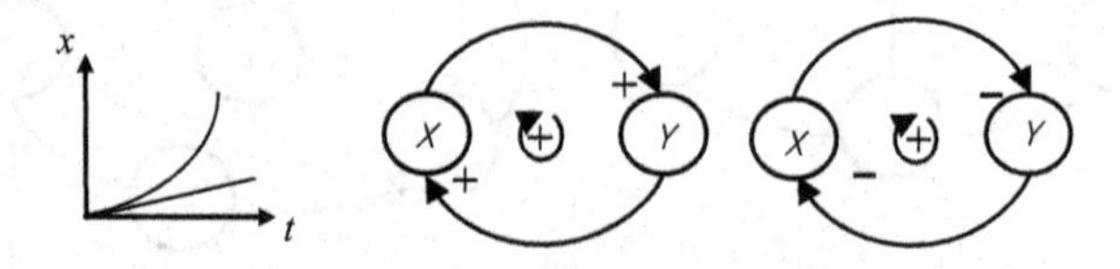

FIGURE 8.6 Positive feedbacks generate nonlinear growth.

(without feedback, only the invested amount) grows linearly over the years, and with compound interest (with feedback, i.e., when calculating the principal amount together with the interest of previous years), the amount of the account grows exponentially. For example, with an annual 10% amount of 10,000 rubles in 56 years will turn into 66,000 rubles with simple interest, and with compound interest, 1,280,000 rubles.

Archetype of regulation. Single *negative* feedback loop (without delays!) produces a monotonic approximation of the value of the parameter X to a given (target) level X_0: if the initial state exceeds the target, the feedback gradually reduces it to the target; if the initial state is below the target, it increases (see Figure 8.7).

Appearance of delays in the communication channels between the elements significantly increases the variety of types of behavior of the system, further generating nonmonotonic (including oscillatory) processes. For example, a delay in the negative feedback loop leads to fluctuations due to delays in the regulation of controlled parameters (see Figure 8.8).

Even in such a relatively simple structure as in Figure 8.8, the nature of the resulting process is highly dependent on the ratio of delays (inertia) of the controlling and controlled systems. (Examples include the fluctuations of the difference between supply and demand on the markets for many products, the sinusoidal character of the trajectory of the car on the road, fluctuations of the regulated parameter around the target value in automatic control systems, temporal variations in the abundance of animal populations of predators and prey, fluctuations in the value of the shares on the stock exchange, etc.)

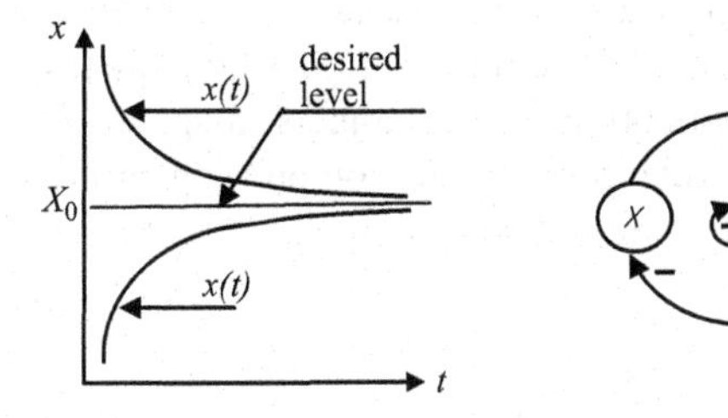

FIGURE 8.7 Regulation without delay.

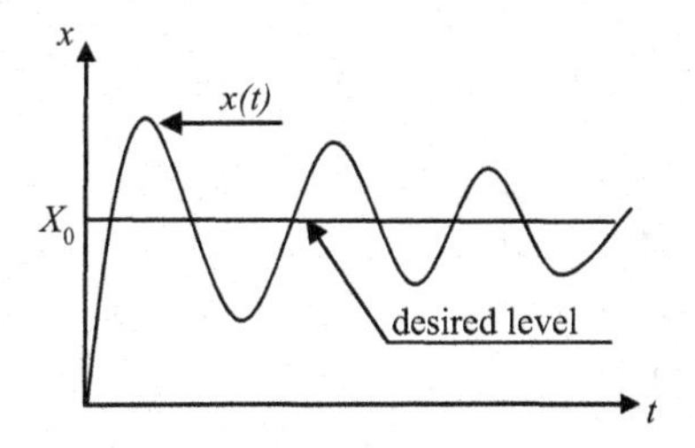

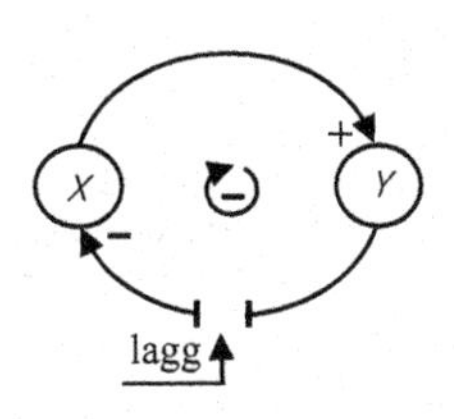

FIGURE 8.8 Regulation with delay.

Let us now consider the archetypes associated with **the two** interacting loops of feedback.

Archetype of Drifting Goals ("Erosion of goals"). The difference between the existing and target states of the system can be reduced in two ways: by bringing reality closer to the goal, and by lowering the level of targeted claims. The last scenario of behavior is this archetype. It is characterized by the gradual lowering of targets leading to the degradation of the system.

Different (internal and external) reasons can force to decrease claims, to force to pay more attention not to that it would be desirable to happen (i.e., the purposes), but to that something did not occur (i.e., to decrease of the purpose). For example, people who reduce their demands are gradually becoming less successful; the country's budget deficit leads to economic crises; reducing investment in the development of the company is actually a form of lowering product quality standards; the weakening of environmental requirements for acceptable levels of pollution reduces the quality of life in it. Structure of feedbacks that causes the archetype is shown in Figure 8.9.

Note that two *negative* feedback loops form one "eight-shaped" *positive* feedback loop (with an even number of weakening links in this chain of links), which leads to the approach of the crisis. Delays cause short-term measures to collapse in the long run. (Clear allegory: plopped in hot water, the frog will instantly jump out of it; but if the water is heated gradually, the frog suffers until it is boiled.).

The "Escalation" Archetype. If there are no delays in the structure of the two *negative* feedback loops, the crisis escalates as an avalanche (due to the eight-shaped *positive* feedback loop). Each of the opposing sides considers the actions of the other side as a threat to themselves, and further strengthens its aggressiveness in response to the increased aggressiveness of the opponent. For example, the escalation of the cold war into a hot war (as at the beginning of the Caribbean crisis), transition of the subjects' dispute to the judicial phase (in particular, divorce proceedings), tough competition in business, rivalry for influence on the chief in bureaucratic structures, arms race, and uncompromising fight against terrorism.

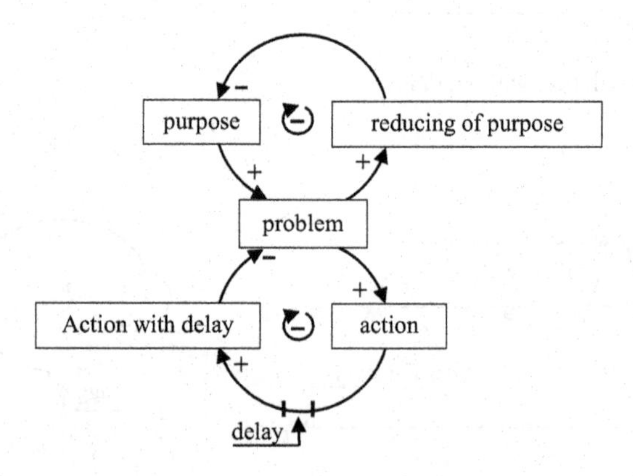

FIGURE 8.9 A problem is a gap between reality and goal.

The structure of the links leading to the escalation events is shown in Figure 8.10.

A cessation of the escalation is possible either in the transition from competition to cooperation as a result of finding a common goal covering the goals of both sides, or in the unilateral termination of the confrontation. Otherwise, the conflict enters an irreconcilable phase.

Archetype "Limits to Growth". A separate class of changes is a *quantitative* increase in a certain characteristic of the system (internal—size, mass, number of elements; or external—force, stress, speed, beauty, superiority, power, etc.). This type of change is called *growth* (of this characteristic).

However, the growth may have a different character. It is found that the accelerating growth is associated with the action of a single positive feedback loop (see "Archetype of growth"). In reality, however, growth never lasts forever; this is due not only to limited resources for growth but also to the fact that other feedback loops and delays in them are involved in the process. This led to the identification of another type of behavior of the growing system — the archetype "Limits of growth."

This archetype is characterized by the fact that the initially accelerated growth of the system begins to slow down and, despite the persistent continuation of successful in the past efforts, completely stops (stagnation occurs) or even goes into a recession stage (crisis, recession). Typical graphs of such a process and a block diagram of the archetype (of two *opposite-valued* feedback loops) are shown in Figure 8.11. Constraints to growth may be of internal and/or external, resource or legal, material or cultural origin. Symptoms of the archetype: At first, "We are growing remarkably;

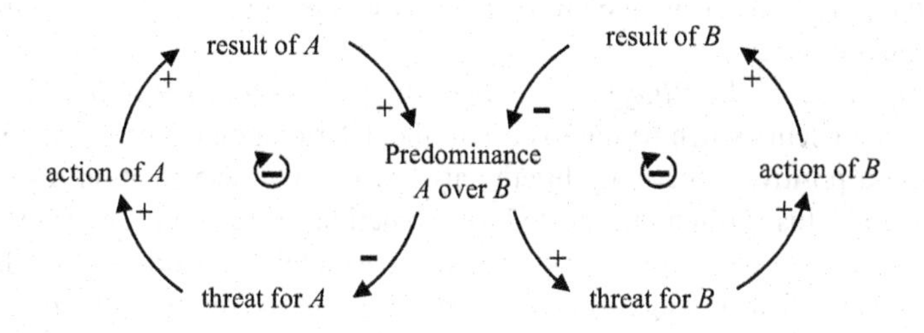

FIGURE 8.10 Archetype "Escalation".

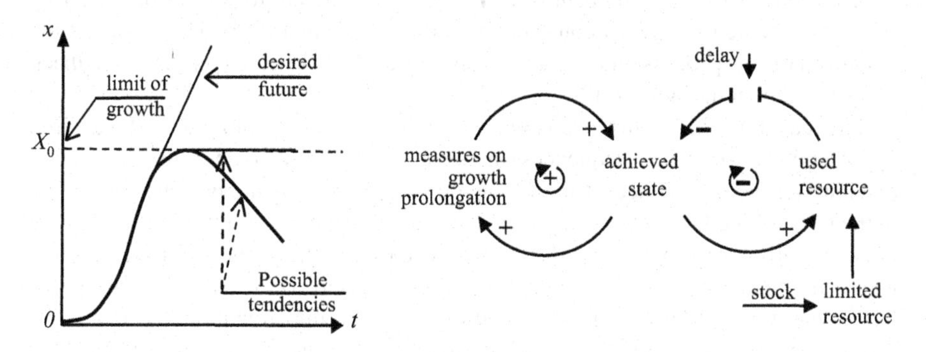

FIGURE 8.11 Archetype "Limits of growth".

we have nothing to worry about." Then, "The problem is, but we will solve it, acting as before." Finally, "Despite all efforts, we are growing more slowly."

Examples of this archetype are observed in the form of a corresponding lifecycle graph (see Figure 3.18) in all types of systems. So is the curve of sales of a new product in economic systems behaves; so is the vital activity of any biological individual changes; such character is the history of all empires; such forecasts of the state of the world economy are given by the "Roman Club" [6]; and "Everything has its end, and its beginning", as the poet said.

The causes of this archetype of behavior can be both outside and within the system and be of a diverse nature. The possibility of postponing the crisis or reaching a stable state is to identify the causes of the slowdown and take measures to counter them.

For example, the annual planned growth of production requires from the company not only a corresponding increase in material resources but also an increase in the number of employees with the necessary skills, and the time of training of new such personnel is much longer than the time of the planned increase in production volumes. If training is not started long before it is needed, a decline in production is inevitable.

Another example is given by the forecasts of the Club of Rome [6]: on the computer model of the dynamics of the world economy, 13 scenarios were calculated, differing in what combination of growth inhibition factors are regulated (extraction of natural resources, pollution of the environment, land productivity, resource-saving technologies, population numbers, volume of industrial production, food production, fresh water resource size, etc.). Twelve of them differ only in the timing of the final global crisis, and only one promises to reach a stable level, with the change of mental models (the paradigm of the meaning of life), the rejection of consumer attitudes to technological progress.

The archetype of the "Ineffective amendments" (Fixes that Fail). This type of situation often occurs when trying to correct the deterioration of the situation in some way, giving a positive effect only briefly, after which the deterioration resumes with renewed vigor. This is often unexpected; but sometimes a person is aware of the negative consequences of the corrective action taken, but either the desire to immediately get relief exceeds the fear of long-term consequences (as in the removal of a heart attack), or he simply does not see another way out (as in the desire to increase sales by increasing investment in marketing due to their reduction in the development of production).

Knowledge of the mechanism of this archetype can help to get out of it, getting into it unexpectedly, or even consciously refrain from entering it. The nature of the dynamics of this process and its causing structure of two different types of feedback loops are shown in Figure 8.12.

The reason for its occurrence is that the action taken has, along with the desired (quick and short-term) consequences (in the negative loop), also unaccountable negative long-term consequences (this is shown in the diagram by the presence of a delay in the positive feedback cycle). In other cases, the cause of falling into this archetype is the initially erroneous acceptance of the symptom of the problem for the problem itself, the means to achieve the goal for the goal itself. (For example, "the problem is that we don't have (or lack) something" or "...to implement modernization."). Overcoming of this archetype (as well as all others) can be carried out with the help of technologies of applied system analysis (see Part II).

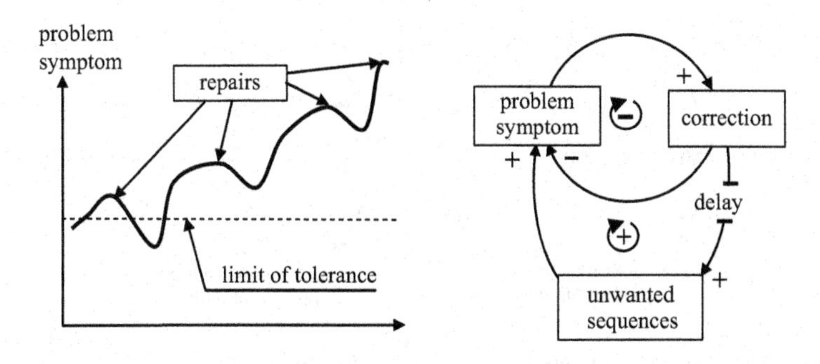

FIGURE 8.12 Archetype "Ineffective repairs".

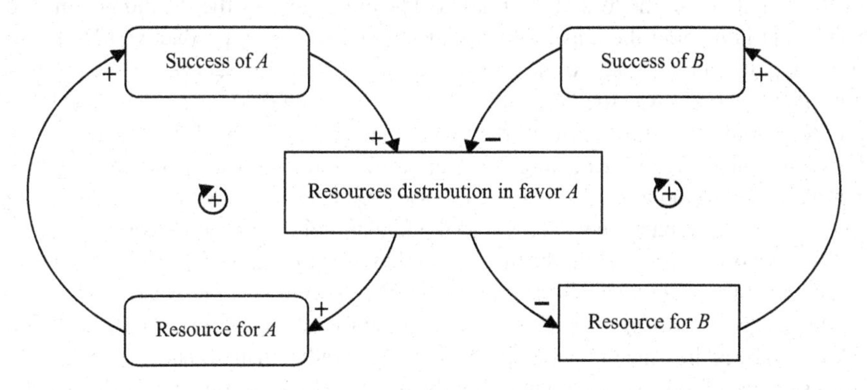

FIGURE 8.13 "Success to successful".

The archetype "Success to the Successful". It is stated that out of two equal systems A and B in the rest, the winner is the one whose initial resource is greater: the first success A is rewarded with an additional resource, which leads to a further increase in its success. The block diagram of this archetype is shown in Figure 8.13.

Examples: Encouraging successful athletes: increased attention to training promising professional athletes; creation of elite educational institutions for gifted children. The only possible order generally accepted only because at first it was followed by the majority: the direction of movement of clock hands, different directions of writing for Europeans (from left to right), Arabs (from right to left), and Japanese (from top to down). The principle of bank finance "Money to money".

However, this archetype has negative aspects. The first is the final emergence of a monopoly position in this area. In some areas (such as sports) this is a desirable goal, in others, it is undesirable (e.g., in the economy it destroys competition, which is why the state takes antitrust legal restrictions). The second was called the "Trap of self-sufficiency" (Competence Trap). Sometimes a high level of previously successful activity is maintained (despite the drop in external demand for it!) only because we are able to do this activity better than others. A good example of this is the history of IBM (Figure 8.14).

Having become a monopolist in the production of large computers (mainframes, MF), the firm, confident in their superiority over the emerging PC, poorly invested

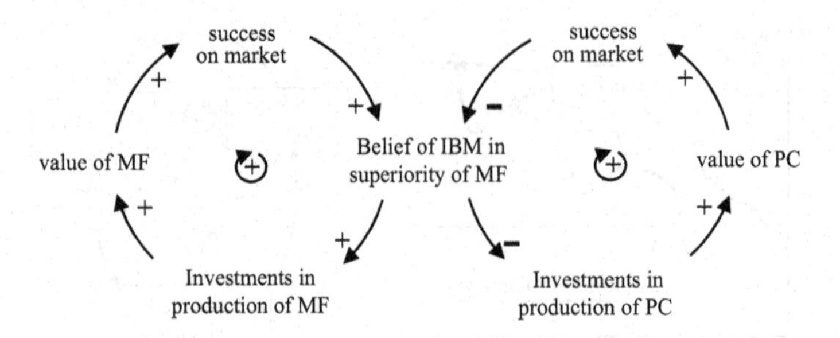

FIGURE 8.14 "Trap of competence" for IBM producing of MF.

in PC, which led to the loss of the market. The emergence of the PC did not change IBM's opinion about the reasons for its success ("users need productive MF"), and the company did not react to the fact that the reason for the success of the product with the consumer was the best ratio of its price and quality. As a result, IBM lost the PC market. In a similar trap of their own competence is a qualified teacher of the university, insisting on increasing the hours of its well-established course, and does not want to develop a new course for themselves. The reason for the "Competence Trap" is the discrepancy between internal and external criteria of success.

Let us now consider some behavioral archetypes that appear in cases when the structure of the system is dominated by more than two loops of feedback.

Archetype "Shifting the Burden". In practice, complex problems are often addressed by addressing the symptom of the problem rather than the cause. A side effect of this choice of the leverage point is that the relief of the symptom creates the illusion of a solution to the problem and distracts from the search for its fundamental solution.

The structure of connections that generates this archetype is shown in Figure 8.15.

Examples are attempts to solve the problem of alcoholism by banning the production and sale of alcohol, the problem of road crashes by increasing fines for traffic violations, the problem of crime by increasing punishment, the problem of stress by taking a dose of alcohol or drug, the problem of paying debt through a new loan, the problem of increasing profits by expanding the traditional segment of the market instead of diversifying production, etc.

One version of this archetype is so widespread and often so harmful that it is worth a separate mention. This is a case of trying to solve the problem only through external efforts, without mobilizing internal capabilities, that is, "Shifting the burden to a third party". There are times when external assistance mitigates the problem to such an extent that the problem carriers themselves lose interest in solving the problem by their own efforts. For example, in some countries, the multichild allowance allows parents to avoid productive work at all; one-time care for genetic patients distracts attention from taking measures to prevent their occurrence

The main principle of external aid management is "Teach them to fish instead of giving them fish." If external assistance is needed, it must be either one-time (and everyone knows this in advance) or aimed at developing the capacity, resources, and infrastructure of the system to succeed on its own in the future, without external assistance.

Archetype "Growth under Underinvestment". This archetype is associated with the situation when the growth of the company reaches a limit that can be surpassed

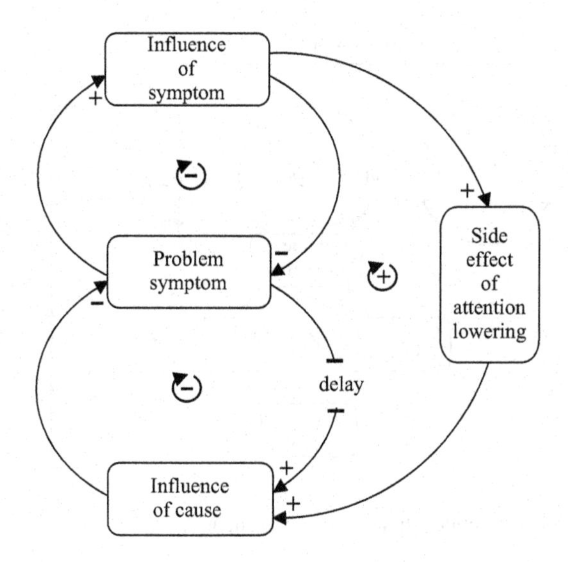

FIGURE 8.15 Archetype "Shift of a burden".

only with additional investment. However, the company can continue to increase production, by reducing the requirements for product quality. This creates the illusion of unnecessary additional investments in production, justifying its further underfunding. But the decline in the quality of goods or services inevitably leads to the displacement of the company from the market. This behavior of the firm follows from the interaction of three feedback loops, combining the archetypes "Limits of growth" and "Erosion of goals" (Figure 8.16).

A symptom of this archetype is that the firm does not attach importance to reducing the quality of its product, but instead blames competitors or its sales department for insufficient sales efforts.

The way to avoid falling into this archetype is to preempt events: if the conjuncture promises real opportunities for growth in demand in the future, the company's strategy should include preparation for this in advance, to begin training the necessary personnel, to create the necessary reserves of resources in a timely manner, to monitor compliance with quality standards, etc.

Archetype "Tragedy of the Commons". This archetype characterizes the situation when all members of the group use a limited, publicly available resource, each satisfying only its own interest. As the resource decreases, everyone tries to increase their consumption, which only accelerates the depletion until the complete exhaustion of the resource.

Examples of this are not only the situation with the consumption of nonrenewable resources but also renewable resources when the rate of consumption exceeds the rate of growth of the resource: traffic jams in cities (resource consumed is traffic capacity of roads); the disappearance of many species of game animals, fish, and plants; pollution of soil, water, and atmosphere by industrial and household human waste; a shortage of qualified personnel in the economy; etc., The structure of the

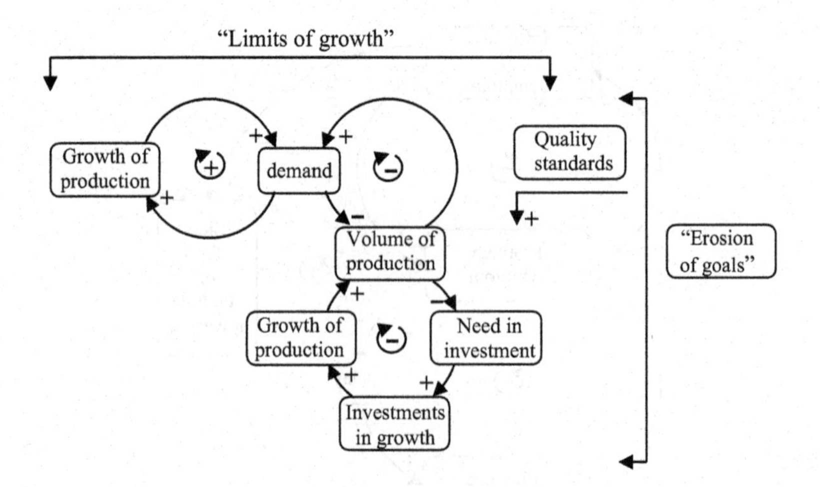

FIGURE 8.16 "Growth under shortage of investments".

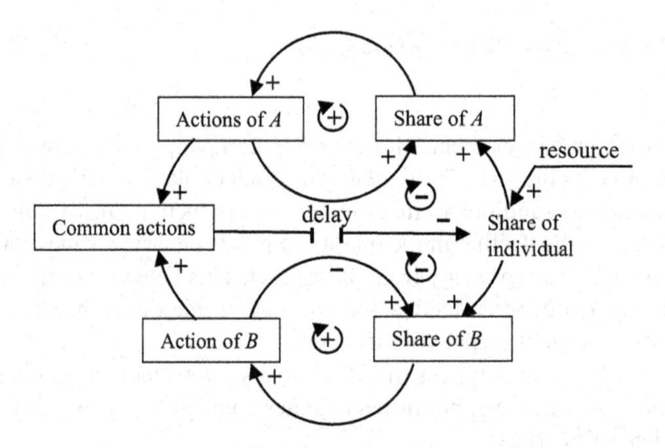

FIGURE 8.17 "Tragedy of the commons".

network of feedback loops and delays that generate the archetype is given (for the case of two consumers of the resource) in Figure 8.17.

To postpone the tragedy (for a nonrenewable resource) or to prevent it (for a renewable one), it is possible only by developing and implementing a strategy of rational regulation of the use of the resource, which harmonizes short-term individual and long-term collective interests.

About other archetypes. We describe the behavior of this system in this environment as a set of processes occurring at the system outputs (i.e., as a set of its *functions*, see Sections 2.1 and 2.2, Part I). Each function is the result of the interaction of all parts of the system (see Section 2.3). However, in what happens at this output, different parts of the system play different roles, which differ not only qualitatively but also quantitatively: the influence of the actions of some parts is more significant than the others (e.g., the absence of one finger did not prevent Yeltsin from playing volleyball perfectly, but excluded the opportunity to play the piano). As a result, in the formation of a specific function, the determining role is played not by the entire network of elements of the system, forming its structural scheme, but only by certain part of it. The structure of this part determines the nature of the process at this output: what and how many feedback loops, and in what combination they participate in this substructure, and determines what is called the *archetype*.

Thus, a process can be considered as an archetype when it interests us more than others in the behavior of the system. To understand why it happens this way, it is enough to identify the features of the substructure that determines this process. If this process is undesirable, you should rebuild this substructure appropriately, creating a new archetype.

A good example of this situation is considered in the case of Marilyn Paul "the Transition from mutual accusations to mutual responsibility" [7]: "When something goes wrong, most often the first question we ask is 'Who is to blame'?. In many organizations, the search for the perpetrators is similar to the conditional reflex. Even those who wish to learn from mistakes are tempted to find guilty ones.

However, there are problems with this: before the threat of accusations, people close, the investigation comes to a standstill, the desire to understand the essence of what is happening sharply narrows. In the atmosphere of the impending accusations, there is a natural desire to hide their mistakes, to hide the true motives, and to blame someone else."

M. Paul [7] established two systems of feedbacks together causing this behavior. First, positive feedback loops increase the difficulty of solving the problem (Figure 8.18). Accusations cause fear, which increases secrecy and reduces the flow of information. Lack of information hinders problem-solving by increasing the number of errors (internal loop). Fear also suppresses the willingness to take risks in the future and the desire to innovate (second cycle).

Second, the "burden-shifting" archetype is involved in the situation, making it difficult to realize that the fault is common, and reducing the ability to make

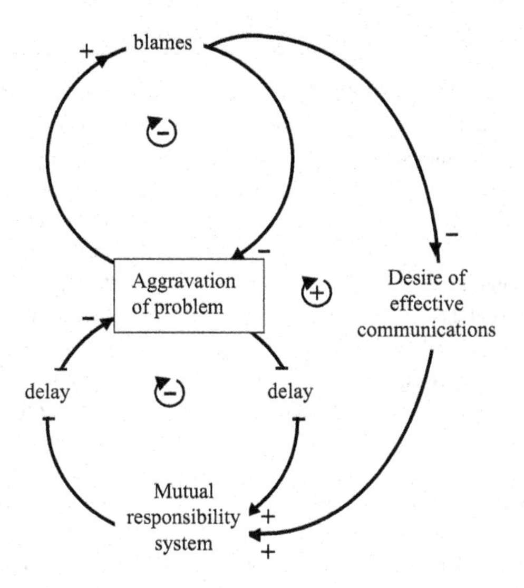

FIGURE 8.18 Inclination to blaming.

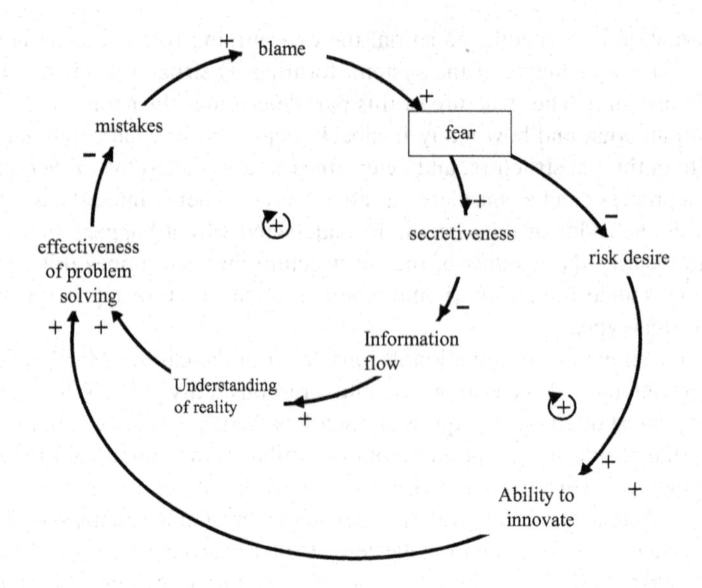

FIGURE 8.19 Increasing of blaming.

common efforts to solve the problem (Figure 8.19). The announcement that someone is to blame alone only briefly creates in majority the illusion of relief of guilt (the upper loop of positive connection). This reduces the desire to share information and communicate effectively (external positive feedback), and undermines the ability to develop mutual responsibility (lower negative connection loop).

Discussing the shift away from mutual accusations to mutual responsibility, M. Paul characterizes the differences between them as follows:

	Charges	Responsibility
Object of systems analysis	**Individual** "Let's find out who is to blame for failure"	**System** "What in the structure of our system increases the likelihood of error and reduces the chances for success"?
Focus	**Personality** "Who did this"? "What did he do wrong"?	**Problem** "What happened to us"?
The intention	**To punish the guilty** You must be punished for your guilt	**To improve the work of the organization** "Let's find out what we have to change in order to achieve the desired results"
The resulting effect	**Stealth, learning disability** "I will hide my mistakes and try not to do anything on my own initiative in future, in order not to risk making a mistake"	**Openness, learning from mistakes** "I want to discuss the mistake, so that we all learn something, and in the future do our job better"

At the end of the consideration of the phenomenon of archetypes of systems behavior, once again we emphasize that it is based on finding the "main reason" for this type of system behavior: "this is the result of the action of such a combination of feedbacks and delays".

However, in reality, the result is influenced not only by this combination but to some extent the other components of the system. The degree of clarity of the archetype depends on the degree of predominance of the influence of its characteristic substructure over the influence of other structures. This explains the fact that the severity of the conflict is different, and the implementation of recommendations in different organizations to overcome the undesirable archetype leads to success to varying degrees.

8.4.3 On Mathematical Modeling of Dynamics of Complex Systems. Synergetics

As in all other areas of knowledge, progress in the study of the nature of systems occurs through the accumulation of experimental information and the development of models of the object of knowledge in the direction of increasing their completeness and accuracy. In the limit, this leads to the desire to build *mathematical* models of systems (see Section 3.4, Part I). It is not surprising that in mathematical modeling of systems physicists advanced the farthest and farthest away: the highest form of knowledge of physical systems are models of theoretical (*mathematical*) physics, generalizing and explaining the results obtained by experimental physics.

The finding by physicists the main system-wide property of nature,—*emergence,* occurred when considering the *dynamics* of physical systems, that is, in the description of natural *changes over time*, the state of a particular physical system. The manifestation of emergence in dynamics is called *synergy* (from the Greek "joint action"), emphasizing the nonlinearity of the common action, that is, not just the sum of particular actions, but the effect of emergence of their interactions.

In physics, it is customary to describe the state of a physical system by determining its observable and *quantitatively* measured characteristics: the values of *constant* (during the experiment) parameters, and the values of *variable* parameters in the course of observation.

For example, a free-swinging pendulum is characterized by constant parameters: the initial amplitude, weight (mass) of the pendulum, the distance between the suspension point and the center of mass, the coefficient of friction on the suspension axis, the coefficient of air resistance to the movement of the pendulum; and variable parameters: the position of the pendulum at each time (deviation from the equilibrium point), and vectors of speed and acceleration of motion at a given time.

At each moment of time, the state of the system is characterized by a specific set of variable values, that is, *coordinates of a point in the state space*: over time, this point moves, forming a certain *trajectory*. The specificity of the nature of this system is manifested in certain stable features of its trajectories in the state space.

In physics, the state space is called *phase space* because for some physical systems different areas of this space correspond to qualitatively different phase states of the system (e.g., for molecular matter in the space "temperature—pressure" is the phase of solid, liquid, gaseous, and plasma states).

It is easy to see that the quantitative modeling of dynamics occurs according to the general scheme with qualitative modeling of static systems. The set of parameters is a model of the system composition. The model of the system structure (i.e., the description of the interactions between the parts) is an equation that relates the parameters to each other. The problem is a description of the inputs (setting the initial state of the system), and its solution is a description of the outputs of the system (together it is a black box model). The solution of the system equation is a *quantitative* description of the synergetic behavior of the system (determination of the trajectory of the emergent state of the system in phase space; the "phase portrait" of the system): each structure corresponds to its inherent archetype of behavior (in synergetics it is an *attractor*).

Already from the qualitative model we know that the archetype is determined by the specific structure of the feedback (positive and negative) links and time delays of resource flows in the links. These features of the system in the mathematical model are displayed by a function that quantitatively describes the relationship between the parameters. For example, the simplest equations of motion describe the relationships between the parameters $x = \{x_1, x_2, \ldots, x_n\}$ and their velocities dx_i/dt:

$$dx_i/dt = V_i(x), \quad i = 1,2,\ldots,n. \tag{8.1}$$

Positive (amplifying) feedbacks are displayed by nonlinear terms in polynomials $V_i(x)$; negative (weakening) by members of attenuation, energy dissipation; and delays by parameters of inertia.

Naturally, different structural schemes of the system, that is, different functions $V(x) = =\{V_i(x)\}$ correspond to different emergent (synergetic) effects. Convergence of dynamic processes to certain specific configurations is interpreted as a process of *self-organization* of the system, which gave reason to talk about synergetics as a theory of self-organization.

In world science, there are scientific schools that develop different specific branches of synergetics: the I. Prigogine's Brussels school of nonequilibrium thermodynamics of macro systems, the Lorenz–Haken school of irreversible microsystems, the Gorky Mandelstam–Andronov school of autowave oscillations, the Zeeman school of the theory of catastrophes, the theory of fractals of Mandelbrot, etc. The specificity of these branches of synergetics is based primarily on the study of different types of differential equations (ordinary and partial derivatives of different orders) and different types of function $V(x)$ (which, incidentally, required a serious development of a number of sections of the mathematics).

However, along with the results specific to each branch, some surprising and fascinating *system-wide* patterns have emerged. Three related topics attracted the most attention.

1. Order and chaos in nature are related: in particular, this is manifested in the fact that in systems described by equations containing only *nonrandom* parameters, at certain values of these parameters, the trajectories become *random*; and in natural *chaotic* processes, there are internal *regularities*.

2. All (both random and chaotic) trajectories of systems in phase space form stable configurations called *attractors*.
3. The structure of natural systems is hierarchical, with uniform rules of organization on all floors of the hierarchy; this feature of nature is called *fractality*.

(All these features are manifested already on the example of the simplest model of system dynamics, Equation (8.1).)

These three aspects of the dynamics of systems together have allowed some progress in understanding the mechanisms of the observed amazing process of self-organization of nature, the development of our cosmic universe; the formation of a discrete spectrum of material entities; the appearance of living organisms; the observed diversity of trends in their changes, one of which led to the emergence of human beings and social systems; the development of diverse interactions between all existing systems (of which we are primarily interested in human interactions with each other and the world around us). Let us briefly discuss the detected features of the dynamics of systems.

8.4.3.1 Order and Chaos in Nature

When performing any of his purposeful actions, a person often faces the fact that what is really happening is not quite or not at all what he expected. Such events are called *random*, and the need to achieve the goal even in the presence of randomness demanded to understand its nature, that is, to build models that allow to operate successfully in conditions of incomplete predictability of events.

The randomness is most evident in gambling; with them began the knowledge of the nature of chance: theory of probability was built (the mathematical axiomatic of which was completed by A.N. Kolmogorov in the middle of the last century) and the game theory based on it. It turned out that the randomness of the events *in this case* is associated with incomplete knowledge of the conditions for their coming true. The probability (measure of possibility) of an event is determined by a certain set of conditions, and if we plan our actions on the basis of inaccurate knowledge or ignorance of any of the conditions, the event (naturally!) occurs in unexpected ways.

A typical example is the dice game. Under the condition of "correctness" (uniformity) of the bone and "honesty" of its throwing, the probability of any points is the same and is equal to 1:6. There is a "probabilistic" anecdote. *Players in the market place bet on different points, and the dice thrower on the six. When he won six times in a row, the players began to beat the thrower, accusing him of fraud. So it was: the bone was shifted center of gravity. But why did they beat him only after the sixth time?*

The concept of random events was soon extended to their time sequence, first, a discrete (statistical series), and then a continuous (random process). It was necessary to introduce the account of variability of conditions (in particular, different variants of dependence of events spaced in time, the concept of ergodicity and stationarity of processes).

In general, this direction of research into the nature of randomness has focused on the cause of the chance, which is connected with what the person interacting with reality knows (and does not know) about it. In this case, random (*for a person!*) events arise because of his ignorance of some objective conditions of their origin.

Another important result of this branch of knowledge of randomness is that even random phenomena are subject to certain laws. All information about stable, nonrandom properties of random events is concentrated in the probability distribution function for such events. For example, in the one-dimensional case, we can talk about the parameters of the position and scale of the distribution function, in the multidimensional case, about the measures of dependence of variables, etc. On the basis of the theory of probability there were developed various applied disciplines — game theory, mathematical statistics, operations research, information theory, theory of economic risks, actuarial mathematics, etc.

But even more important progress in the knowledge of the nature of chance and the relation of chaos and order was the discovery that randomness is inherent in nature itself, regardless of whether the person is involved in the events with his knowledge.

Actually, this fact was discovered at the beginning of the century in quantum mechanics with its uncertainty principle and wave functions, but at first it looked more like a feature of the mathematical model of the microcosm. But later it turned out that even in strictly determined systems described by exact equations that do not contain any random parameters or functions can occur typically random events. To describe this phenomenon, more general concepts of chaos and stochastic events, reversibility and stability of systems were introduced; various quantitative characteristics of stochastic uncertainty were introduced: entropy, moments of distributions, measures of the amount and value of information. (Detailed critical reviews of this section of dynamic systems theory are contained in books [8–10].) There were found out two (for now?) sources of objective stochastic events in nature.

The first was discovered at the beginning of the century by Poincaré in the form of a class of *unstable* systems whose phase trajectories diverge sharply at the slightest differences in the initial conditions. It turned out to be impossible to obtain analytically the exact solution of the equations of such ("nonintegral") systems, but it was possible to carry out a qualitative analysis of their dynamics in computer experiments.

In particular, the sensational popularity of this phenomenon was obtained after the publication of the case that occurred in weather forecasting by solving a multidimensional system of equations on the computer linking previously observed in the area of weather data with possible upcoming data. Having received the forecast with very accurate initial data (up to six decimal places), and leaving for lunch, the meteorologist E. Lorentz decided just in case to recalculate the forecast to speed up the calculations by setting the same data with an accuracy of only three characters. Imagine his surprise when he returned from lunch and saw the second forecast drastically different from that of the first! The investigation revealed that the decision of some nonrandom (!) equations is extremely sensitive to initial conditions: a very small change in initial conditions can lead to very different results.

This systemic effect was found not only in mathematical models but also in real nature, and even got the figurative name "butterfly effect". (Scientific folklore connects this name with two sources. The first is a joke: "If a butterfly flaps its wings at the mouth of the Amazon river, a typhoon will break off the coast of Japan." The second is the story of the science fiction writer Ray Bradbury about how a tourist traveling by a time machine in the era of dinosaurs, and warned about the categorical ban to change there anything, accidentally crushed a butterfly on the path. When he returned home, he saw America as a completely different society.)

The high sensitivity of a complex system to small changes in the initial conditions became the subject of the study of its stability factors, giving rise to the corresponding branch of the mathematical theory of differential equations.

Along with this, another manifestation of objective randomness in the behavior of systems described by completely deterministic equations was discovered: the presence of stochastic "attractors" in their phase space.

8.4.3.2 Attractors

Consideration of the equations of dynamics of systems found that the behavior of real systems occurs in an orderly, regular manner. It turned out that nature does not allow any behavior, but only one that belongs to a finite set of possible phase portraits of the system. When describing the behavior of the system in the form of a trajectory in phase space, this is expressed in the existence of a *discrete and finite* set of its subdomains, within one of which the trajectory of the system behavior is located. Different subdomains differ in the specific values of the factors that determine this situation. For example, the behavior of water in the phase space "temperature and pressure" is determined by four subdomains corresponding to the solid, liquid, gaseous, and plasma states of water.

This feature of nature was manifested in the study of mathematical models of system dynamics. For example, it was found that for the systems described by Equation (8.1), there are only four types of their trajectories in the phase space "coordinates and velocities". These types are interpreted as "centers of attraction" ("attractors") for real trajectories of the system:

- *point attractors* — stable behavior in time, displayed by a particular point in the phase space;
- *cyclic attractors* — rhythmic switching from one type of behavior to another and back (usually these are oscillations between two types of action, but there may be situations when these types are more);
- *toroidal attractors* — the case when the trajectories of the system behavior in the phase space are chaotically different, but remain within (or even just on the surface) the *torus*, the volume figure, the "tube" formed when the circle moves along a closed trajectory (like a boublik, which, however, can have not only the correct form);
- *strange attractors* — "bundles" of random trajectories occupying a limited part of the phase space, but, unlike toroidal attractors, having irregular and random boundaries.

The fact that such a typology of the behavior of natural systems has a system-wide character indicates the existence of four phase states of physical substances (solid, liquid, gaseous, and plasma), and the presence of four styles of managerial behavior of managers, discovered by R. Ackoff [1]: *reactive*, passive (*inactive*), *proactive*, and *interactive* management. Examples of cyclic attractors can serve not only the oscillations of the pendulum or Buridan's donkey but also switch of the flaps of a heavy aircraft only from minimum to maximum and back (for optimum smoothness of the trajectory), as well as periodic elections of political parties in two-party democracies (one party is for freedom, capitalism, the other for equality — socialism; these two ideas are incompatible at the same time, but can be realized alternately, upon the occurrence of a preponderance of disadvantages over advantages of one another). The transition to another attractor can also serve as a change of laminar flow to turbulent with increasing water pressure in the tap or with increasing the speed of the ship.

Further consideration of mathematical models of systems dynamics revealed that the number of attractors may differ from four and may be very large, depending on the number of degrees of freedom and the ratio between numbers of positive and negative nonlinear indicators in the equation (we can see a close analogy of attractors with different archetypes, the implementation of which depends on the relationship between positive and negative feedback loops and delays in different loops).

Attractors are forms of stable states of real systems, the final states to which the evolutionary dynamics of systems is directed. Therefore, it is natural to interpret them as objective goals of systems under given conditions (the relationship between subjective and objective goals was discussed in Section 2.3 in Part I), and the process of approaching them as a process of *evolutionary self-organization*. The behavior of the system in the zone of instability (near the boundary separating the zones of attraction of different attractors like a watershed of neighboring rivers) accompanied by a jump-like switching of the system from one attractor to another (such an event in synergetics is called *bifurcation* or *catastrophe*) is interpreted as a *revolutionary self-organization* of the system.

The discovery and study of attractors led to the realization of a number of natural peculiarities (laws) of systems. Two of them will emphasize particularly.

The first is the discovery that the randomness of real events can have not only a subjective reason (ignorance or incorrect knowledge of the subject of some objective conditions of the origin of the event) but also be quite objective (i.e., not at all related to the existence of people) feature of the behavior of nonrandomly organized systems. In the study of the mechanism of chaos in the system described by the equation without random components, it turned out that chaos often occurs with repeated doubling of frequencies in the spectrum of the phase trajectory. This is due to the hierarchical self-reproduction of the structure of the system, which leads to a feature of nature called *fractality* (which we will briefly discuss later).

The second is the establishment of the fact of *hierarchical discreteness* of the set of all manifestations of nature. In the system dynamics it is a discrete (and finite!) set of attractors of the system. In system statics, it is the discreteness of the elements, structures, and archetypes. In the physical microcosm, it is the discreteness of the set of elementary particles, atoms and molecules, and the discreteness (quantization) of their energy states. In the macrocosm, the discreteness of cosmic bodies, stars, planetary systems, galaxies. In the world of living systems, it is a hierarchical discreteness of individuals, their species, families, populations, ecological communities. In social systems, types of social formations.

8.4.3.3 Fractals

The modern concept of *fractals*, a very specific feature of the structure of natural systems, gradually formed as a result of the efforts of many natural scientists and philosophers to understand the principles of *self-organization* (origin, growth, and development) of natural systems.

The fact that fractals are inherent in nature was discovered and understood gradually, in parts, from different viewpoints on reality. One aspect of fractals — the *self-similarity* of natural structures, or *wide-scale invariance* — was noticed in ancient times by Eastern thinkers. Buddhist monks from ancient times learned the world, thinking alone about themselves ("contemplating his own navel"). British missionaries in colonial times discovered in the culture of the African Zulu tribe the concept of umuntu — a person who is aware of his continuity with the community, experiencing suffering if others suffer. The Zulus have a reserved wisdom: umuntu ngumuntu ngabantu ("everyone is part of everyone"). Many of the thinkers of all time have come to realize this property of natural systems. Three hundred years ago, Leibniz wrote: "Any part of matter can be represented as a garden full of plants, or a pond full of fish. But every branch of the plant, every member of the animal, every drop of its juices is again the same garden or the same pond." And the outstanding psychologist of the last century Carl Jung saw the similarity between the structures of the universe and the human psyche: "Our psyche is arranged like the universe; what happens in the macrocosm, occurs in the smallest, most subjective corners of the psyche."

Both in science and art there were displays of the phenomenon of cascade (hierarchical) structure of reality. Paintings by Leonardo da Vinci "Avalanche" (Figure 8.21) and by Katsushik Hokusai "Big wave" (Figure 8.20) illustrate the phenomenon of self-similarity: all the smaller parts of the whole are organized at all levels in the same way. Folk art also reflects this: remember the nested design of toy "Russian dolls"; songs like "I, you, he, she, together — the whole country"; the British in the

FIGURE 8.20 An artistic illustration of hierarchical structure of reality.

FIGURE 8.21 Another artistic illustration of reality structure.

course of the saying such as “I am who I am because of everyone”, the Australians — “I am, you are, we are Australia”.

This aspect of the self-similarity of the structure of natural systems can be called a hierarchical system (bearing in mind that the system is sometimes non-hierarchical). Features of hierarchy (see Section 8.4) generate stability, efficiency, and mutability (discreteness of development branches) observed in natural systems, which largely explains the mechanisms of self-organization and evolution in nature, and at the same time the reasons for the frequent preference for hierarchical organization of management in technical and social systems.

But the real breakthrough in understanding this feature of systems, which ended with the formation (and naming!) the concept of *fractality* occurred in the development of a *mathematical* model of geometric configurations. To this day, the outlines of many flat and three-dimensional shapes are usually described using continuous smooth lines and surfaces in terms of Euclid geometry. Drawings of all artificially created by us technical objects (tools, devices, buildings, structures) look this way. But these objects are *artificial* systems. Natural systems, however, have a different geometry. As Benoit Mandelbrot writes in his pioneering book *The Fractal Geometry of Nature* (Freeman, San-Francisco, 1977) [11,12]:

“Clouds are not spheres, mountains are not cones, the outlines of the shore are not circles, and the bark is not smooth, and lightning does not spread in a straight line. Nature shows us not just a high degree, but a completely different level of complexity. The number of different scales of lengths in structures is always infinite. The existence of these structures challenges us in the form of the difficult task of studying those forms that Euclid discarded as formless — the task of studying the morphology of the amorphous.”

Long before Mandelbrot, in several sections of mathematics (in geometry, set theory, logic) abstract objects were already built, so strikingly different from the usual regular objects that they were called “monsters”, “paradoxes”. Examples of geometric objects of this type are the Peano curve (Figure 8.22):

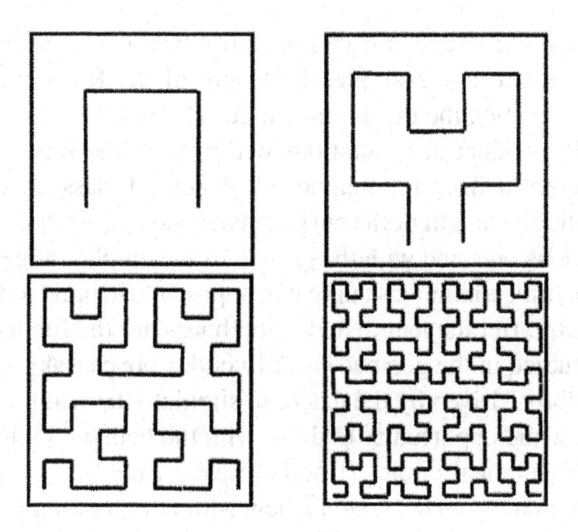

FIGURE 8.22 Four generations of the Peano curve.

and Serpinsky's triangles (Figure 8.23);

in the set theory, such an object is the Cantor dust (Figure 8.24);

in logic — the paradox of Zeno ("Achilles and the turtle") and the paradox of the liar ("I lie").

Some of the salient features of such objects can be clearly seen in the construction process of the Koch curve introduced by the Swedish mathematician Helge von Koch in 1904. We begin with a straight segment of unit length ($n = 0$). Divide it into three equal (1/3) parts, discard the middle part and replace it with a polyline of two links in length 1/3 so that the middle part is the base of an equilateral triangle. The result was a broken line of four parts with a total length of 4/3 ("first generation", $n = 1$).

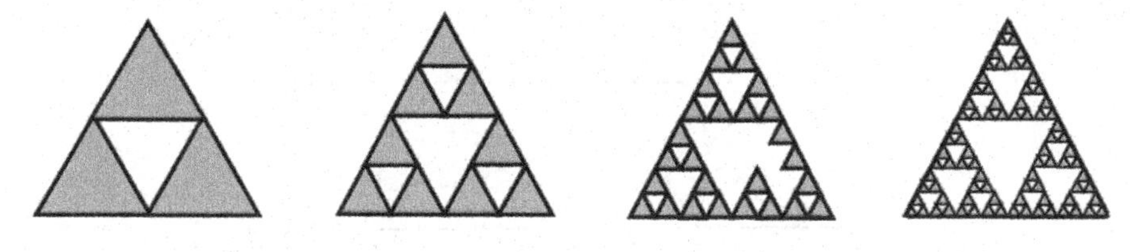

FIGURE 8.23 Four generations of Serpinsky's triangles.

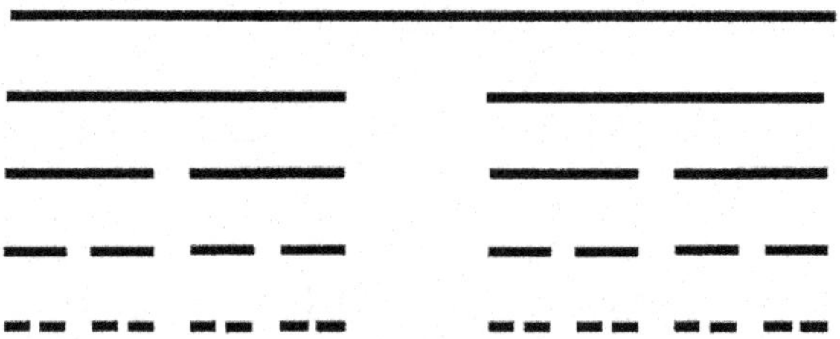

FIGURE 8.24 Five generations of the Cantor dust.

The next generation will succeed if this operation is repeated with each segment of the previous generation. For example, the length of the Koch curve of the second generation is already 16/9, the third—64/27, etc. (Figure 8.25).

With an infinite product of generations will process the Koch curve. It has such properties that derive it from the number of objects of classical geometry: (1) this curve has no length: the length of the curve of each next generation is greater than the length of the previous one, and with the growth of the number of generations tends to infinity; (2) to this limit continuous curve it is impossible to build a tangent: each point is an inflection point. But the length and smoothness are the fundamental properties of objects to be studied in the geometries of Euclid, Lobachevsky, and Riemann.

Benoit Mandelbrot, driven by the desire to simulate the real properties of natural configurations, saw the opportunity to do so with the help of "monsters" that caused such a stir in mathematical circles (Bourbaki even called them "pathological phenomenon in mathematics"). He created a new geometry, calling it *fractal geometry*. Fractals, objects of new geometry, allowed to build formal models of "formless" natural chaos. Mandelbrot so justified the introduction of a new term: "I created the term ***fractal*** from the Latin adjective *fractus*. The corresponding Latin verb *frangere* means to break, interrupt, i.e. to create irregular fragments. It therefore has (suitable for us!) a meaning additional to the term 'fragmented': fractus is also 'irregular'. (…) The combination of a '*fractal set*' will be defined strictly, but a combination of '*natural fractal*' will be given freely — for the definition of natural examples that are useful to represent using fractal sets".

The last phrase speaks of a clear understanding that the mathematical fractal is an abstract model of real natural chaos, more useful than the model of regular geometry: fractals are fundamentally different from the traditional geometric forms such as point, line, and smooth surface.

The construction of fractal geometry led to a tremendous breakthrough in the knowledge of nature finally completing the transition to the paradigm of

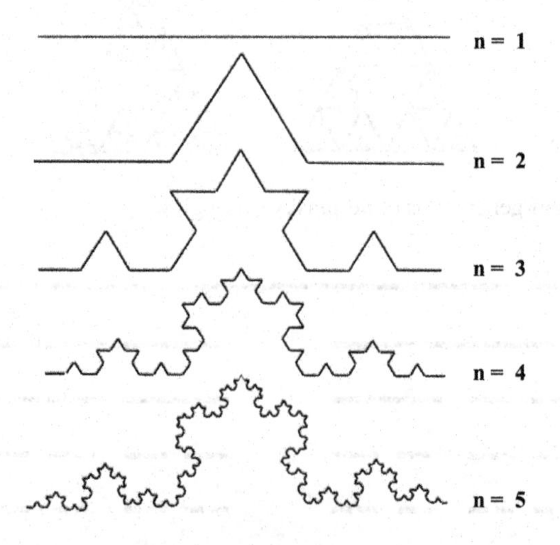

FIGURE 8.25 First five generations of the Koch curve.

discreteness of all things. It was found a striking similarity of mathematical geometric fractals with natural (continuous!) geometric configurations (e.g., the trajectory of lightning, the curves of the river, the outlines of clouds, mountain ranges, coastlines of islands and continents), which sometimes seem unpredictable chaotic, random, and sometimes very complex, but amazingly harmonious (the shape of snowflakes, the sound of music, the external forms of every living creature, a variety of plant and animal species, the shape of space objects, such as planetary systems and galaxies).

This fact led to the understanding that all natural phenomena in their original essence have a discrete nature. Particular aspects of this universal feature of nature were discovered earlier: by explaining the nature of matter, physicists have built a *quantum* model of the microcosm; chemists have detected a *discontinuity* in the world of chemical elements (periodic table); biologists — "mutafacient" forms of life (*discreteness* of species and the impossibility of their crossing); sociologists have found that the society consists not only of *individual* subjects, but also of *individual* cells (family, ethnic group, nation, group, organization, community).

However, it has now become clear that not only spatial but also *any* qualitative continuous (in some respect) natural phenomena are inherently discrete. For example, the continuous trajectories of dynamical systems are implemented in the form of a discrete set of *attractors*; continuous growth of the system occurring through accumulation of resource from the environment is implemented in the form of separation of discrete *parts* of the system (branches of plants, the deposits in mineral deposits, settlements in the inhabited territory, the private capital in the market economy, etc.).

This internal, initial discreteness of even continuous manifestations of nature is represented in their mathematical modeling by a hierarchical and self-similar structure of fractals: generations of continuous fractal sets are constructed discretely, iteratively, by the same relations for all generations. A good example is the set of Julia and Mandelbrot points (x, y) on the plane. Both sets are built recursively by the transition from the k-th point to the $(k+1)$-th point using the functions F and G:

$$x(k+1) = F\left[x(k),\ y(k),\ p\right],$$

$$y(k+1) = G\left[x(k), y(k), q\right],$$

where p and q are some parameters of the constructed set. The functions F and G may be different but necessarily nonlinear.

If on the plane consisting of all pairs (x, y), we consider the limit position of the initial point z at k tending to infinity at fixed p and q, then we obtain sets of Julia. In particular, if on the complex plane, at $z = x+iy$, recurrent iterations z, and the nonlinear function $F(z, c) = z\times z+c$,—then appear attractors (!). When $c = 0$ there are three attractors: (1) $|z| < 1$, the attractor is the point (0,0); (2) $|z| = 1$, the attractor is a circle of radius 1; and (3) $|z| > 1$, the attractor is infinity. The fourth attractor, other than zero, occurs when $c \neq 0$.

If we trace the trajectory of the point when the parameters p and q change, then various images of an extremely complex multicomponent configuration, called Mandelbrot sets, appear on the plane. For example, in Figure 8.26, the configurations

FIGURE 8.26 The Mandelbrot set in different scales.

of the more and more small sizes of the same set are given. Each next image (from *a* to *e*) is a rectangle circled and enlarged fragment of the previous one.

Fundamental difference between fractal geometry and classical geometry also manifested itself in attempts to describe fractals in terms of Euclidean geometry. It was impossible to obtain fractal configurations by analytical technique of integral and differential calculus based on continuity and smoothness; fractal images can be constructed only algorithmically, using discrete iterative procedures on a computer. The description of some properties of a geometric object through the concept of its dimension also encountered the inapplicability of this classical concept to fractal objects.

The dimension of a geometric object is the number of coordinates that can be used to determine the position of each point of the object. Dimension of a straight line — one; dimension of a smooth flat figure — two; volume — three. To describe the position of the points of fractal (irregular, tortuous, rugged) objects is not enough coordinates of the smooth set in which they are placed. A fractal point is no longer a point, but not yet a line; a fractal line or figure on a plane is no longer flat, but not yet volumetric (Figure 8.26). To describe this property of fractals, the definition of dimension (Bezikovich–Hausdorff) coinciding with the integer dimension of smooth sets and characterizing the degree of difference between a fractal set and a smooth one is proposed. This concept of dimension is based on the idea of covering the considered set with elements $h(\delta) = \delta^d$.

Suppose we have a set of points in certain Euclidean space. Let's cover this set densely in turn with straight line segments ($d = 1$), squares ($d = 2$), cubes ($d = 3$). The measure of M_d of a set is defined as $M_d = \Sigma\delta^d$. The measure is a general concept for length ($d = 1$), area ($d = 2$), volume ($d = 3$), and is computable also for noninteger d. At $\delta \to 0$, the measure M_d is zero, or infinite if the dimension of the measured set and the d-dimension of the measure do not coincide. For example, when covering a flat figure with straight lines M_d is infinite, when covering it with cubes, the measure tends to zero (the volume of the plane is zero), and when $d = 2$, the measure of the flat figure tends to the final value (the area of the figure).

If you cover the Koch curve with segments ($d = 1$), its measure (length) tends to infinity; when covered with squares ($d = 2$), its measure (area) tends to zero. There is a value of $1 < d < 2$ at which the measure changes its value from zero to infinity; it is an estimate (1,2628…) of the dimension of this fractal: this is what is meant when it is said that "the Koch curve is no longer a line on the plane, but not yet a volume figure". It is not easy to determine the fractional dimension value of a particular fractal. Estimates of the length of the coastline give different dimensions for the coasts of Australia, Germany, the west coast of Britain, Portugal [last].

This speaks not about the "strangeness and complexity" of nature but about the limitations of our knowledge about it and the imperfection of our models of reality. Projecting ideas of fractality not only onto description of the geometric qualities of reality revealed that the fractal character is inherent in many other (and may be in all?) aspects of the organization of nature: the attractors in the dynamics of systems archetypes in the static structures of systems, uniformity in the probability measure of distribution of sample values of the observed quantities, diversion of living species, self-similarity of branching plants, historical analogies, etc.

Mandelbrot realized his intention to reveal fractality not only of quantitative manifestations of nature but also of phenomena evaluated qualitatively, in weak measuring scales, by introducing a strictly mathematical concept of "fractal set" and a qualitative concept of "natural fractal". The corresponding distinction of dynamic processes is made in the form of constructing a qualitative analog of the numerical concept of "phase space": the coordinates fixed in weak scales form a "phase space".

The study of natural phenomena revealed the universality of the properties of fractals, but at the same time showed that the mathematical fractal is an abstract, approximate model of the organization of nature: real systems are not fractal, but only *fractal-like*. First of all, this is manifested in the fact that the number of levels in real hierarchical systems is not infinite, as in mathematical fractals, but is limited: the hierarchy of self-similarity begins to develop from the lower level of *qualitatively specific* systems and ends with the appearance of lower-level systems of the next qualitatively new, higher level of development. For example, the hierarchy of the physical microcosm ends with the appearance of chemical molecules, the biological hierarchy begins with the appearance of a living cell, the hierarchy of social systems begins with the elemental communities of individuals (family, pack, primitive community). In addition, the likeness of future generations in reality is not their identity.

Awareness of the phenomenon of fractality was a major breakthrough in the knowledge of how nature is "arranged".

Questions and Tasks

1. How does a person achieve the one-function (narrow specialization) of an artificial system with the natural multifunctionality of any object?
2. How can you explain the different effectiveness of the control action on the material, energy, and information "leverage points" of the system?
3. Formulate definitions of positive and negative feedbacks.
4. What explains the various behaviors of systems (archetypes)?
5. What features of the systems dynamics lead to the formation of fractals?

9 Elements of Complexity Theory

9.1 FORMATION OF THE CONCEPT OF COMPLEXITY. RANGE OF DIFFICULTIES

All sciences accumulate a variety of information about the world and about humans as its element. However, the science of systems pays special attention to the *interactions* of people with their environment, that is, the *cognition* and *transformation* of reality by humans. In the study of human activity to change the environment in accordance with their goals, a special place is occupied by clarifying the conditions for successfully achieving the goal, determining the causes of difficulty (*complexity*!) in achieving the goal, as well as finding ways to overcome, bypass, or at least reducing complexity. These problems are the central subject of applied systems analysis.[1]

The transition from the Age of Machines to the Age of Systems in the activities of mankind (see Chapter 1) takes place in the form of developing management practices in a continuously changing environment, with increasingly encountered difficulties in achieving the goals. During the research regarding ways to overcome these difficulties, the experience of "difficult" situations was accumulated and generalized; gradually, the understanding of the essence of the phenomenon of complexity in human activity in the surrounding reality increased. Understandably, the explanation of the incomprehensible is reduced to the representation of the accumulated information in the form of three basic models: models of the "black box", composition, and structure of the considered system. All the difficulties in our practice arise from the "difficulties" lurking in the resulting models, which we further use in the development, adoption, and execution of our management decisions. We now understand that the source of all difficulties lies in the fact that our models, that is, ideas about the reality that we want to transfer into the desired state, are somewhat different from the reality; and that these differences can relate to the different elements of any of the models, as a result of which complexity can have different nature, vary in degree of complexity, as well as in quality characteristics.

[1] Note that since in English the meaning of the word *analysis* is reduced to *reductionism*, that is, consideration of only the composition (sometimes, structure) of the system, and the Russian term *applied systems analysis* includes a synthetic consideration of the system (i.e., the black box model), in English system terminology, it corresponds to the terms *systems thinking* and *design thinking* (project thinking).

9.2 CLASSIFICATION OF THE COMPLEXITY TYPES

The simplest model of diversity is classification (see Section 3.5, Part II). Like all models, classification has a purpose: different classifications are built for different purposes. Consider the most common classification of types of complexity designed to "localize" the reasons for this type of complexity (finding "lever points"), followed by the definition of ways to reduce the complexity of this type (fully or to the minimum level possible in this case).

9.2.1 Classification According to the Degree of Objective Complexity in the Behavior of the Controlled Object

Study of systems dynamics has revealed the discreteness of the types of behavior of such systems, and it was found to be subject to completely deterministic differential equations of the second order: their trajectories in the phase space converge to one of the four possible configurations (point, cyclic, toroidal, and strange *attractors*, see the section "Attractors" in Section 4.4.3). The real sensation was the discovery of random trajectories inside toroidal and strange attractors. This makes the remote future of the system unpredictable, making it very difficult to manage.

The discovery that nature itself limits the filling of the phase space with trajectories of the dynamic system only within a limited area of the attractor, and the set of attractors is discrete and finite, with the attractors themselves differing in a growing degree of chaotic trajectories, was one of the three great achievements of the 20th century in the knowledge of nature (the first two include the theory of relativity and quantum mechanics). The feeling that this is not only a physical regularity of striking beauty but a manifestation of the universal law, which is fair to all forms of organization of the material world, was confirmed by the consideration of chemical, biological, and social systems: everywhere there are similar features of the dynamics of the state of the system.

In particular, in the theory of management of social systems, there are four types of systems: *simple, complicated, complex,* and *chaotic*, in order of increasing *complexity* of management (due to the growing uncertainty of the future). For the first three types of complexity, methods were developed to overcome the difficulties (control algorithms). Naturally, the algorithm for overcoming a particular difficulty is associated with the possibility of neutralizing a specific cause of complexity — the lack of some resources (substance, energy, information) necessary for successfully achieving the goal.

Different types of control correspond to different reasons of complexity (see Section 4.3 in Part I):

- simple system management (all necessary resources are available) —program management;
- management of a *complex* system (with lack of information about the managed system) — trial and error (Note that here the term "complex" is used in a narrow sense when the reason for the complexity is only a lack of awareness about the managed system);

- parameter control, or *regulation* (for small, compensable differences between the model and the real state of the environment);
- management structure, *reorganization* (in case of large deviations, non-compensable due to the lack of material resources);
- *target* management (if the current target is unattainable);
- management of *large systems* (with lack of time to find the optimal solution). (Note that here the terms "big" and "small" also have a special meaning which differs from everyday);
- management of the *uncertainty in the ultimate goal* (if information is unavailable about the ultimate goal, but you have confidence in the existence of the best state) — heuristic (revolutionary) and empirical (evolutionary) approaches.

However, in relation to chaotic (objectively stochastic) systems, the question of control is different: the laws of nature are not subject to us, and we cannot change the natural randomness of events. It remains only to adapt to the unexpected changes taking place around, like fishermen in a storm, or pilots caught in a zone of turbulence, or a surfer, maneuvering on his board on the slope of a steep wave. As D. Meadows puts it, chaotic systems cannot be controlled, but they can be danced with [13].

System thinking offers a set of techniques for managing social systems that have entered a chaotic phase of their lifecycle. These techniques are based on the use of information about the objective laws of natural development, including fractality, change of attractors, self-organization, etc. (see Chapter 3). However, experts warn [14] that due to the presence of people's consciousness and free will, the development of global information technology and other factors, chaos in social systems is much more complicated than physical chaos, which requires caution in the practical use of the results of the mathematical theory of chaos.

9.2.2 Complexity Classification of Types of Models of the Managed System

Targeted impact on reality is planned on the basis of a working model of the system undergoing transformation. Which of the three basic models (black box, composition model, structure model) or a combination of them should serve as a working model in a particular case depends on the characteristics of the problem and the problem situation. However, when constructing any particular model, errors may creep into it (see Section 2.1, Part I), or the constructed model may be inconvenient to use. All this leads to difficulties and creates *complexities* in work. Again, we are faced with different types of complexities that require different approaches. Let's consider the classification of difficulties that may arise in each of the basic models.

9.2.2.1 The Complexity Caused by the Large Dimension of the Composition Model

In some cases, developing a successful solution to the problem requires information about all elements of the problem situation, that is, a fairly complete model of the composition of the system. The importance of this requirement is related to one of the basic laws of cybernetics — *the Law of Requisite Variety* proposed by Ashby. The variety of the system is the number of its possible states. Ashby's law claims that problem-free management is possible only if the diversity of the controlling system is not less than the diversity of the controlled system. Often complexity is due to the lack of diversity of the system composition model. There are systems consisting of a very large number of elements, and working with their composition model becomes extremely time-consuming and difficult. The reason for the complexity in such cases is the lack of available modeling resources to complete the processing of all information at the required time. In computational mathematics, this difficulty is called the "curse of dimension"; in control practice, such systems are called "large".

A good example of this situation is the delay of three to four years of compilation by the USSR State Planning Committee of the annual input–output balance among the millions of produced and consumed products. This alone (in addition to other difficulties) sharply reduced the effectiveness of rigidly centralized operational management of Soviet Union's national economy.

There are two options for controlling a large system (see Section 4.3, Part I): demanding physical acceleration of simulation (increasing computing power, parallel simulation techniques), and speeding up the simulation by simplifying the models (intentional reduction in the number of considered variables, linear approximation of nonlinear dependencies, etc.), that is, conscious low-quality, but timely decisions.

As a measure of the complexity of large systems, A. N. Kolmogorov proposed to use the length (the number of bits in the program) of the algorithm that completely describes the composition of the system.

9.2.2.2 The Complexity Caused by the Inadequacy of the Structure Model

It is known that the most characteristic properties of the system, the emergent properties, are determined by the structure of the system (see Section 2.3, Part I). Therefore, in the disclosure of the complexity of the behavior of the system it is very important to identify the features of the structure that determine this emergent property of the system. This is clearly manifested in determining the causes of specific types of behavior of systems — archetypes: the reasons for this lie in the specific combinations of feedback loops in the structure (see Section 3.4.2). This is an example of studying the system, that is, building a model of its structure. In cases when the working model of the controlled system includes its *structure model*, control difficulties arise if the model contains errors of any of the four types (because the structure is a certain combination of inputs and outputs of all elements of the system). The problem with this type of complexity is the adoption of measures to prevent errors in the structure model (see Section 2.1).

9.2.2.3 The Complexity Caused by Incomplete Information in the Combined Operating Model of the Managed System

The working model of the system can be a combination of the composition and the structure models of the managed system. If the working model lacks any information necessary to achieve the goal, the system falls into the category of "complex". This term has a specific conditional meaning: it does not mean the property of the object, but the relationship between the controlling subject and the controlled object, it is the *complexity of ignorance*. The same object can be "complex" for one subject and "simple" for another: they have models of this object of varying degrees of adequacy. (For many, the control of the TV is a simple matter, but its repair difficult.)

Complex systems of this type are managed by trial and error method (see Section 4.3, Part I). This algorithm extracts a new piece of information about the system from the result of each subsequent ("trial") control step, adds it to the existing model, and then plans the next control action according to the improved model. Thus, the method with each step increases the adequacy of the model, reduces the complexity of the system, and increases management efficiency. A measure of this type of complexity can be the number of accomplished cycles of algorithm, which help achieve a satisfactory approximation of the goal.

9.2.2.4 The Complexity Produced by Probabilistic Uncertainty

The transition from the above static models to dynamic ones leads to the introduction of another class of complexity — the complexity of the control of random processes. We emphasize the difference between this type of systems from chaotic systems, the management of which is reduced to "dance" with them, that is, to attempts to adapt to the peculiarities of a particular chaotic implementation — by identifying a pattern, the links between the events, and attempts to "fit" into their flow in our own interests. But in this case, we are talking about the possibility of using information not only about the observed realization of a random process but also about the whole ensemble of its realizations, and even about the possibility to influence the characteristics of the process itself.

Such possibilities appear if the randomness of the trajectories is not completely chaotic, but organized and limited by a certain framework, as is the case in a strange attractor, but not only in it. The "limitation" is that such a randomness is fully characterized by the probability distribution function over the possible realizations of a random phenomenon (event or process). Moreover, the constraints on the process include the conditions of stationarity (statistical stability of each realization) and ergodicity (statistical homogeneity of all realizations).

The complexity of working with such an object is caused by the *uncertainty* of predicting its behavior. The observed or controlled parameters of a random object can be chosen as appropriate characteristics of its distribution, for example, the parameters of the position or scale of the distribution and various measures of connection between variables (correlation, regression); as a convenient measure of uncertainty distribution, C. Shannon proposed *entropy*. Any desired numerical characteristic of a random object can be expressed by some *functional of the distribution function*, analytically calculated or statistically estimated value of which is used for the object. The possibility of controlling a random process, that is, a purposeful change of some of its characteristics, is associated with the possibility of changing its probability distribution. Here, the dual nature of the concept of *probability* as a measure of uncertainty of the outcome of a random experiment comes to the fore.

The probability of outcome is defined as a measure of its possibility to occur *under a given set of conditions*. This set of conditions includes the laws of the stochastic nature of the phenomena (which determine the *objective* component of probability);

however, the level of uncertainty (and hence the value of the probability) is also affected by the *degree of knowledge* by the subject about objective conditions, which gives rise to the *subjective* component of probability. (A good example of controlling a random object due to the subjective part of the probability is given by a sharper in card or dice games; improving the quality of the weather forecast is carried out by taking into account additional factors affecting the weather; etc.).

Different degrees of reliability of knowledge about probability distribution dictate the need to extract the necessary information from the same data set in different ways. Special algorithms of experimental data processing have been developed for different levels of a priori information about the distribution. Mathematical statistics consists of four sections:

1. classical (parametric) statistics based on the assumption that the distribution function is *known* up to a finite number of parameters;
2. nonparametric statistics, assuming that the observations are subject to distribution, the functional form of which is *unknown*;
3. robust statistics considering cases where the distribution function is *known approximately*: the real function is located in some neighborhood of a given function;
4. semi-parametric statistics, assuming that observations belong to a parametric family, *with random parameters.*

In accordance with this, the "Rules of statistical safety" when working with random objects are developed. Methods have also been developed to incorporate any additional (collateral) information into statistical data analysis procedures.

9.2.2.5 The Complexity Associated with "Vague" Uncertainty

Classification of difficulties will be incomplete if we do not mention another type of complexity that often occurs in human practice. The above types of difficulties are typical for the cognitive or transformative activity of an individual subject, including it under the conditions of uncertainty associated with the *probabilistic* nature of the object of activity. However, uncertainty in the working model can be not only *random.*

In human practice, very often there are situations where certain activities must be carried out by several persons, together and in concert. This means that each of them should have its own working model of the situation in which it plans its actions; however, to coordinate the actions of all participants, it is necessary that the different models contain the same information about the common situation. (The most ancient example of the failure of the collective project is given in the biblical parable about the failure of the construction of the tower of Babel because of the incompatibility of the languages of the builders.)

Therefore, models are built in the languages of the problem situation configurator (see Section 5.5, Part II). If the corresponding language is professional, that is, it describes the situation accurately for all participants in the collective work (e.g., the languages of professional mathematicians, engineers, or repairers), then there are no difficulties in the work. But often (especially in the management of social systems) the configurator includes spoken language, with its characteristic ambiguity of the meaning of words, especially evaluative words that express the gradation of quality in weak measuring scales. This leads to complications in the joint work of subjects by giving them different meanings of the same evaluative word.

Vagueness and uncertainty in the meaning of words in the natural language generate another specific type of complexity, which has a purely *subjective origin* (since the ability to evaluate anything has only the subject). Almost all disagreements and conflicts that complicate the joint activities of people arise from the vagueness of the classification terms in the natural language ("What is good and what is bad?").

Mathematical tools to describe this type of complexity were proposed by L. Zadeh in the form of the theory of *fuzzy sets*. The basic idea of this theory was the introduction of the *membership function* $\mu_{class}(x)$, a numerical measure for the degree of confidence of the subject in the membership of the estimated object x to the fuzzy class of objects.

The value range of the membership function ranges from 0 ("not exactly belongs to") to 1 ("surely belongs"). Unlike a precise classification, in a fuzzy classification, there are no clear boundaries between classes, the functions of belonging to them can overlap, and this object can simultaneously belong to several classes (with different degrees of confidence).

For example, the fuzzy classification of numbers into three classes of "small", "medium", and "large" from the viewpoint of one subject can be determined by the dotted membership functions in Figure 9.1, and from the viewpoint of another by a continuous one.

In their joint work, it is necessary to develop solutions that take into account the views of both participants. To carry out the required information processing, the rules for calculating the membership function of specific combinations of different judgments are defined. For example, the "or" operation (U), $\mu^1 U^2 \mu^2_{class}(x) = \max[\mu_{1class}(x), \mu_{2class}(x)]$; for the operation "and" ($\bigcap$) — $\mu^{1...2}_{class}(x) = \min[\mu_{1\,class}(x), \mu_{2\,class}(x)]$, etc.

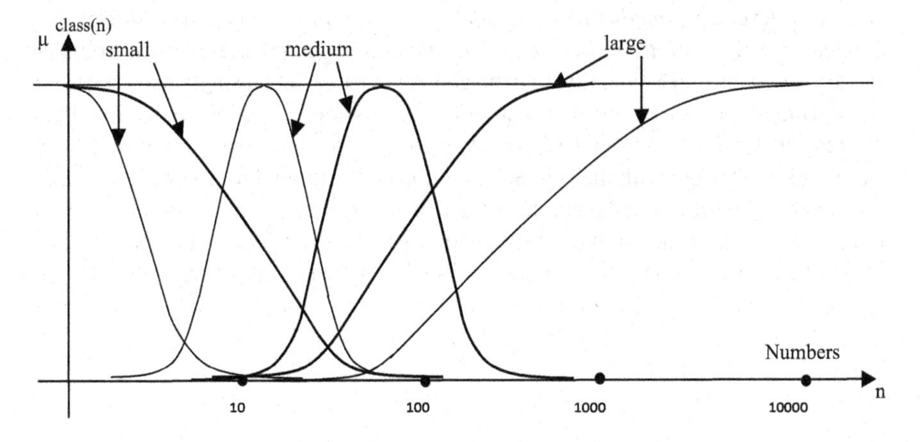

FIGURE 9.1 Fuzzy classes of numbers.

Management decisions are formed with the help of fuzzy logic in the automated processing of vague initial information about the controlled system. It is worth noting that practitioners sometimes wrongly interpret the value of the membership function as probability. This is a gross error: the properties of the membership function and the probability distribution function differ both in meaning and in the technique of their mathematical transformations.

Questions and Tasks

1. What is the meaning of the evaluative word "complex"?
2. Which characteristics of the system are chosen as classification criteria for the different classifications?

9.3 CONCLUSION

In conclusion, it should be noted once again that classifications, like all models, describe reality only of interest to us and only approximately, with satisfying accuracy. Reality is always more diverse than our ideas about it. Therefore, the real difficulties that lie in wait for a person in his interactions with the environment may have a character that does not meet the characteristics of any particular type of difficulty from the above classification.

Quite often, the cause of a particular difficulty encountered in work is the simultaneous action of several types of complexity (e.g., statistics combine objective and subjective uncertainty). Usually, in such cases, a hybrid algorithm is built to overcome the combined difficulties, constructing a suitable combination of ways to work with these particular difficulties (which is not an easy task).

In practice, however, more often we have to deal with such natural situations, the complexity of which is not fully covered by our rational models. In the professional language of systemology, such problems are called "soft", "chaotic", and "confusing, (wicked)". Although some of their partial subproblems can be formalized (e.g., by methods of *operations research*), and for others several heuristic methods of consideration have been developed (*methodology of soft systems, brainstorming, synectics, project thinking, leverage points, search for connections and patterns*, etc.), their complete solution is still beyond rationality. In management practice, increasingly used conscious use of irrational recourses of the brain—the subconscious, intuition, abduction. In an effort to get closer to fulfilling the *law of Ashby's necessary diversity*, the complexity of our brain is contrasted with the complexity of the nature around us. In one respect, this has already been achieved: the number of possible combinations of states of all neurons in the brain exceeds the number of elementary particles in the universe; however, the possibilities of the brain in the search of proper combinations are unknown

Part III References

1. A. Leopold. *Round River.* New York: Oxford University Press, 1993.
2. R. L. Ackoff, F.E. Emery. *On Purposeful Systems.* New York: Aldine, 1972.
3. J. F. Forrester. *Industrial Dynamics.* New York: Pegasus, 1961.
4. R. L. Ackoff, J. Strumpfer. Terrorism: A Systemic View. In F. P. Tarasenko *Problems of Governance*, vol.1, issue 2, Tomsk: TSU Publishing House, 2009.
5. D. Meadows. *Leverage Points. Places to Intervene in a System.* New York: The Sustainability Institute, 1997.
6. D. Meadows et al. *Beyond the Limits.* Past Mills, VT: Chelsea Green Publishing Company, 1992.
7. M. Paul. *Moving from Blame to Accountability.* Westford, MA: Pegasus Communications, Inc., Publications, PG07, 1997.
8. S. D. Haitun. *From the Ergodic Hypothesis to the Fractal Picture of the World.* Москва: URSS, 2007.
9. S. D. Haitun. *Human Phenomenon.* Москва: URSS, 2008.
10. D. S. Chernavsky. *Synergetics and Information.* Москва: URSS, 2009.
11. B. Mandelbrot. *The Fractal Geometry of Nature.* San-Francisco, CA: Freeman, 1977.
12. V. V. Tarasenko. *The Fractal Logic.* Москва: URSS, 2009.
13. D. Meadows. Dancing with Systems. *Problems of Governance*, Vol. 1, 2009, pp. 46–55.
14. M. C. Jackson. *Systems Thinking.* New York: Wiley, 2009.

Index